Many of us have been convinced that being glued to our devices will make us feel more connected, help us prove that we're the best activists, or cure our fear of missing out. In *The Unplugged Hours*, Hannah Brencher lovingly reminds us that what we may actually be missing out on is the full, vibrant life that's right in front of us—one where we can gather, create, learn, and find what we've been looking for all along.

DANIELLE COKE BALFOUR, illustrator; activist; author,
A Heart on Fire: 100 Meditations on Loving Your Neighbors Well

In a world that often leaves us feeling burned out and spread too thin, *The Unplugged Hours* is a refreshing reminder that the life we crave is within our grasp. Hannah comes alongside us with a gentle yet powerful invitation to put down the phone, embrace stillness in the pockets of the day, and get back to the wonder of the present moment. With honesty and grace, this book offers a much-needed guide to a simpler, steadier way of life.

EMILY LEY, author; founder, Simplified

The Unplugged Hours by Hannah Brencher is a timeless message. In her brilliant work, she invites us to step away from pervasive digital norms and rediscover the richness of life offline. Clear, compassionate, and practical, Brencher emphasizes the importance of rest, renewal, and finding our lives again—focusing on what truly matters and fostering a deeper, more meaningful existence away from the screen and in the true gift of life—on the other side of the screen.

ALEXANDRA HOOVER, speaker; ministry leader;
author, *Eyes Up* and *Without Wavering*

The Unplugged Hours is an invitation to rediscover the magic of mundane moments. If reaching for your phone is a h̶a̶b̶i̶t̶ and disconnecting feels difficult, this beautiful work by Hannah to the life right in front of you. As nah's thoughtful stories about learning to balance tech while building an intentional life—one unplugged hour at a time.

KAIT TOMLIN, national bestselling author;
dating coach; founder, Heart of Dating

If your soul is craving more of the peace and presence of God, you probably need to unplug, and Hannah's book is a beautifully crafted road map that will lovingly guide you through the process. When you release yourself from the obligation to constantly consume and give yourself permission to embrace a slower pace of living, you will begin to see God breathing new life into barren places. This book will challenge you to look up and around instead of down. Take the challenge and watch restoration come one unplugged hour at a time.

LASHONDA BROWN, tech educator;
founder, Bootstrap Biz Advice

This book is for the one yearning to get their life back. In *The Unplugged Hours*, Hannah Brencher weaves together faith-filled wisdom, personal experiences, and practical steps forward in a grace-filled guide to the present, joyful life you've been longing for.

NATALIE FRANKE, *USA Today* bestselling author, *Gutsy*

The Unplugged Hours will challenge you to put your phone down and find your life! In a world filled with guilt-inducing advice about social media and dopamine dangers, Hannah offers a heartfelt guide to cultivating presence in a digital world. With wisdom, vulnerability, and grace, Hannah's refreshing writing changed my view on the gift of time. Through poignant personal insights and a transformative challenge of one thousand device-free hours, she inspired me to rediscover the beauty of real connections, embrace vulnerability, and find solace in the quiet moments of life. A must-read for those longing to break free from digital distractions and embrace the richness of authentic human connection.

LARA CASEY ISAACSON, entrepreneur;
author, *Make it Happen* and *Cultivate*

This book will change your life. Hannah Brencher has long been one of my favorite writers and thought leaders. She always seems to know what to say and how to say it in a way that rewires something deep within you that's felt stuck for years. That's exactly what she's done with this book. With wisdom and warmth, and with the courage that comes with going first, she walks us out of a life of hurried emptiness and into a life of deep connection—with God, our people, and ourselves. I can't wait to follow her lead and see what's waiting for me in the unplugged hours.

STEPHANIE MAY WILSON, author, *Create a Life You Love*; host, the *Girls Night* podcast

In a world that conditions us to scroll for more, *The Unplugged Hours* reminds us to scan for presence. This is a must-have book for anyone who feels like a screen has become a barrier to their reality.

JESS EKSTROM, speaker; author; founder, Mic Drop Workshop

Hannah Brencher's *The Unplugged Hours* completely transformed my perspective on the use of technology in my home and family culture. If you're a young mother or part of a family eager to be more intentional in your daily life, this book will pivot every paradigm you may have. Hannah beautifully illustrates how putting away our phones can create more space for joy and connection, especially with our children, without an ounce of shame. By putting her words into action through simple, actionable steps, we have seen more laughter and meaningful moments in our home. If you've ever felt tethered to your phone and longed for more authentic interactions of true overflow with your family, I can't recommend *The Unplugged Hours* enough. It's a must-read for anyone ready to reclaim their time and attention.

RACHEL AWTREY, author; host, *Real Talk with Rachel Awtrey*

If you're ready to take your eyes off your phone and put them back on your life, start here. One part pep talk, one part love letter, this book is Hannah Brencher at her finest. Sharing hard-won wisdom in her signature sister-friend voice, Hannah compels readers to swap a plugged-in life for one of presence. Featuring zero guilt and heaps of grace, *The Unplugged Hours* is the invitation I didn't even know I needed, and one I'll surely return to again and again.

ASHLEE GADD, author, *Create Anyway: The Joy of Pursuing Creativity in the Margins of Motherhood*

In *The Unplugged Hours*, Hannah Brencher writes us all one massive permission slip—she empowers us to put down our phones, our laptops, and our remotes for the sake of wholeness in order that we might come back to life. She does this not with superiority but with vulnerability, as one who has gone before. Hannah takes us by the hand, brings us alongside, and ardently reminds us that our lives are worth paying attention to and being present for.

TABITHA PANARISO, therapist; author, *Loyal in His Love*

Hannah Brencher's *The Unplugged Hours* is a transformative guide to reclaiming your life from digital overload. Each chapter is filled with wisdom and practical advice for cultivating a life of presence. In a culture of always on and constantly drained, everyone needs this book.

ALLI WORTHINGTON, author; business coach; entrepreneur

the
unplugged
hours

the
unplugged
hours

CULTIVATING A LIFE OF
PRESENCE IN A DIGITALLY
CONNECTED WORLD

HANNAH BRENCHER

ZONDERVAN
BOOKS

ZONDERVAN BOOKS

The Unplugged Hours
Copyright © 2024 by Hannah Brencher

Published in Grand Rapids, Michigan, by Zondervan. Zondervan is a registered trademark of The Zondervan Corporation, L.L.C., a wholly owned subsidiary of HarperCollins Christian Publishing, Inc.

Requests for information should be addressed to customercare@harpercollins.com.

Zondervan titles may be purchased in bulk for educational, business, fundraising, or sales promotional use. For information, please email SpecialMarkets@Zondervan.com.

ISBN 978-0-310-36773-4 (audio)

Library of Congress Cataloging-in-Publication Data

Names: Brencher, Hannah, author.
Title: The unplugged hours : cultivating a life of presence in a digitally connected world / Hannah Brencher.
Description: Grand Rapids, Michigan : Zondervan Books, [2024]
Identifiers: LCCN 2024012349 (print) | LCCN 2024012350 (ebook) | ISBN 9780310367703 (trade paperback) | ISBN 9780310367727 (ebook)
Subjects: LCSH: Internet addiction—Religious aspects—Christianity. | Habit breaking—Religious aspects—Christianity. | Cell phones—United States—Social aspects. | Brencher, Hannah. | Christian women—United States—Biography. | Christian women—Conduct of life. | Christian life—United States. | BISAC: RELIGION / Christian Living / Family & Relationships | TECHNOLOGY & ENGINEERING / Social Aspects
Classification: LCC BV4596.I57 B74 2024 (print) | LCC BV4596.I57 (ebook) | DDC 616.85/84—dc23 /eng/20240520
LC record available at https://lccn.loc.gov/2024012349
LC ebook record available at https://lccn.loc.gov/2024012350

The author is represented by Mackenzie Brady Watson of Stuart Krichevsky Literary Agency.

Cover design: Faceout Studio, Jeff Miller
Interior design: Sara Colley

Printed in the United States of America

24 25 26 27 28 LBC 5 4 3 2 1

Mom,

You've given me so many gifts throughout my lifetime,
but the greatest of them has been your presence.
Thank you for teaching me how to pass the gift forward.

CONTENTS

Part Three: Rhythms

Part Four: And Beyond

BEFORE
ANYTHING ELSE

When I unplugged for the first time, I thought that was all I was doing. I'd presented myself with a challenge: one thousand unplugged hours in one year. I hoped I'd complete the challenge, maybe become more disciplined with my phone habits, and move on with the rest of my life.

But early in the unplugging, I realized this was not just a simple challenge. This was not a onetime experiment. This was me—hour by hour—taking back a life that had slowly become less present, less awake, and less vibrant over time.

For many years, I allowed the constant connectivity that came through my phone (and other electronic devices) to be one of the most essential parts of my life. My phone was the thing I woke up to, the thing I went to bed with, the thing I checked at all hours of the day. I didn't realize how much it had taken from me until I started intentionally unplugging and slowly putting technology back in its place.

The book you're holding in your hands today is an ongoing story—the culmination of one thousand unplugged hours and well beyond. It's a story of learning to find a balance between being plugged in and being powered down. It's a story about trading an "always on" mentality that left me frazzled and lacking peace for a chance to cultivate a lifestyle of profound presence.

What you will *not* find in the pages that follow is a proposal that we move toward an extreme lifestyle of using flip phones, deleting all our apps, disavowing technology, and moving off-grid to start a commune where we communicate primarily through carrier pigeons and smoke signals. Some days I think I'd make a great leader of an unplugged commune, but then my mom calls on FaceTime and she and my daughter get to have a breakfast date, even though they're a thousand miles apart. Or a voice memo pops up in my text messages, and I click to hear a friend praying for me, speaking blessings over my day. Or, after over a decade, I reconnect with a high school friend on Facebook, and I'm suddenly thankful these little portals of connection exist.

Tech has many precious and important gifts to offer us. It's the reason we have so many modern-day medical advances. Tech is how many businesses form and how good movements spread. Most of us are tethered to technology in some form or fashion as we work or go about daily life. It's just the way the world is—but I believe there are ways we can learn to live beyond the call to constant connectivity and find a path of balance that is better and more life-giving.

While I reference the one thousand unplugged hours challenge throughout the book, many of the themes tucked within these pages were learned and honed beyond the one-thousand-hour mark. Throughout this journey, I've learned that the best things—the most unshakable lessons—take time to root themselves in our souls. My takeaways were not instant or overnight, but the result of a steady, hour-by-hour practice of powering down over many months—and now years.

The more I've pressed in over the course of the journey, the more I've realized it was not just a call to be unplugged. It was a call to be relentlessly present in my life and the lives of others. It was a call to recover lost parts of myself. It was a call to create rather than just consume. It was a call to see the blessing in the mundane bits of daily life. It was a call to step back into connection, wonder, and devotion while breaking free of the constant stream of hurriedness that left me feeling anxious and restless. It was a call to relentlessly check into my life—and keep checking in. I pray you will experience some of the same calls and even more on your own journey.

Speaking of that journey, what are you waiting for? You don't have to reach the end of this story to start doing the one thing you likely need to do if you've picked up a book about the beauty of unplugging, or if you're curious about what powering down could do for your life. You can begin today with a single small step: Turn off your phone. Right now, if you can.

The other day, a close friend (who has heard me talk about the unplugged hours ad nauseam at this point) texted and told me she was ready to start unplugging. She had told herself she'd start in the new year—like many of us do when we have a goal we're a little afraid to pursue—but knew she couldn't wait until then. She had to start immediately. She shared a vulnerable truth with me in that conversation: She was afraid to turn off her phone because as long as she had her phone in her hand, she could convince herself she wasn't deeply lonely.

Her fear is a common one. When your phone has become your primary source of connection, the thought of powering down is vulnerable and scary. It can feel wildly uncomfortable at the starting line. But if you've reached a point where you feel like connectivity is stealing something from you, I pray this will be your gentle push: *It might be time to shift.*

I'll give you the same encouragement I offered my friend: Start with just the first hour. Start that small.

Think about something you've been wanting to do for a long time, something micro you keep pushing off because of email, mindless scrolling, or yet another binge-worthy show on the latest streaming app. Maybe you've been wanting to bake again. Or you've wanted to dig into a new devotional but keep getting distracted. Perhaps you want to read the book that's been sitting on your shelf for months or pull that puzzle down from the hallway closet and attempt to put it together. Maybe you want to sit outside for a few uninterrupted moments and take in the sounds. Whatever it is, set your intention for the hour. Power down your phone or place it out of sight. Then do that thing you've been wanting to do.

After one hour, turn your phone back on. Everything will still be there. The world will still be pushing forward at a rapid pace. Emails will still be streaming in. People will still be posting the same stories and arguing in the comments section. You'll have missed out on nothing, but gained something back instead: a piece of your time, a tiny sliver of your life. This is how your story will unfold—slowly, one hour at a time.

When you power down that first time, you'll think that's all you're doing. But then you'll find yourself noticing a sunset for the first time in ages, or you'll reconnect with an old friend over coffee, or you'll decide to write the first sentence of that novel you stopped dreaming about but haven't yet entirely forgotten. And a piece of yourself will feel like it's returning to life. You'll realize then that what you're doing is so much more than unplugging.

As time goes on, the unplugged hours will be a teacher to you all on their own, if you let them. They may show you how to navigate boredom, or they may be the spark that brings back the creativity that used to teem inside you at all hours of the day. The unplugged hours may help you discover contentment with your life as it already is, or they may help you see, with fresh eyes, some things that have needed shifting for a very long time. They may lead you to connect more deeply with yourself and others than you thought possible. The unplugged hours might bring you to a place of stillness—of being able to sit alone with yourself—or they may propel you deeper into prayer, or seeking, or questioning. I can't tell you what will happen when you power down, but I know it will be worth it. It is always worth it to release ourselves from the noise of constant connectivity and see what awaits us in the silence.

It has been my ongoing hope that the stories and revelations within this book would nudge you softly into building your own collection of unplugged hours. So go ahead. Turn off your phone and find your life. It's waiting for you to step back in.

As you move through this book, I pray my words will meet you at all the right junctures, like a trusted companion walking through life beside you. I wrote these words imagining us sitting across from one another—sipping coffee, sharing stories, our phones tucked away. I thought about the encouragement you might need to hear. I considered the doubts and fears that might be brewing in your mind right now. I wondered what questions you might ask along the way and what remarkable things you might learn about yourself, about others, about God, or simply about life in the process. I thought about the things I wish I could go back and tell myself at various points in my own unplugged hours. And then I gathered all these thoughts together and wrote them down.

Welcome to the unplugged hours, friend.

THE 1,000 UNPLUGGED HOURS CHALLENGE

Welcome to the 1,000 Unplugged Hours Challenge! The premise is simple: Unplug for one thousand hours over the course of one year.

I've included a tracker at the end of this section so you can start unplugging right away and tracking your progress as you go. If you want to download and print the official tracker, or if you want to opt for the 100 Unplugged Hours Challenge instead (it's a little more bite-sized and perfect as a monthlong challenge), you'll find both printable trackers at www.HannahBrencher.com/Unplugged.

If you prefer to track in some other way, or to make your own tracker, go for it! This is your journey and I'm simply here to cheer you on. To help you get started, here are some lessons and encouragement I've learned through my own journey of unplugging.

DEFINE "UNPLUGGED"

First things first: It's essential to define what "unplugged" means for you. We all live different lives with different demands. This challenge is meant to be restorative, not restrictive, and I think that means setting clear yet flexible boundaries for ourselves.

For me, an unplugged hour means several things: my phone is away, I'm not on email or browsing the internet, I'm not on social media or watching a show, and I'm not consuming any form of digital media—not even podcasts or audiobooks. For you, an unplugged hour might involve a different set of parameters. I encourage you to think about what it means for you to be unplugged in this season of your life and to build your boundaries intentionally. You know yourself better than anyone else does.

One question people often ask before taking on this challenge is, "What if someone needs me or there's an emergency?" It's important to note that I'm not saying you *must* power down. In some seasons of life, powering down might not be feasible or realistic. But in my own life, I've found that if I need to be accessible but want to take a break from my phone, I can leave it on but put it out of sight or in a box. The box I use isn't fancy or new; it's a repurposed IKEA tin. I keep it in my office for when I want to do focused, unplugged work. If my daughter is at school or I'm waiting on an important call, I keep the ringer or vibration on, but otherwise my phone is out of sight. Other options include keeping your phone in your bag when you're out and about, placing it in the center console of your car while driving, or utilizing airplane mode.

MAKE A PLAN

Having a solid plan to unplug is everything. As with nearly all good things, we need to be intentional about making space for what matters most to us.

If you struggle to get the hours in at first, take a look at your day or week and ask yourself, "For what parts of this day or week do I want to be unplugged?" Maybe you want to stop bringing phones to the dinner table. Perhaps you want to be fully powered down for your morning quiet time. It could be that you want to practice a digital Sabbath for an entire day on the weekend. Whatever the case, identify some pockets of time when you'd like to explore being device-free. It's okay and even a kindness to yourself to set boundaries—times and places that are "unplugged zones." The more you do, the easier it will be to unplug seamlessly and make these unplugged hours a lasting rhythm in your daily life.

ENJOY THE PROCESS

When I first started powering down, I was a meticulous tracker of the hours. But as time went on, my focus became less about completing a challenge and more about cultivating a lifestyle. If I could go back to the beginning, I'd tell myself: Enjoy the process.

One challenge participant told me that when she realized she wouldn't hit one thousand hours in a year, she quit. It's not uncommon to have a pass-fail mindset when we take on a challenge, but let's scrap that mindset for a second and consider a few questions instead: *What would you gain by unplugging more often? What is not working in your life that you want to change? How might unplugging help you do that one thing you've been wanting to do?*

Don't make the metrics your focus. Ultimately, the challenge isn't about meeting a quota of daily unplugged hours; it's about building an unhurried, in-the-moment life.

Every hour counts. Even if you get to the one-year mark and you're at hour nine hundred instead of one thousand, that's still nine hundred hours you reclaimed, and you likely will have seen much evidence that the unplugged hours are creating an impact in your daily life! In my corner of the world, we choose to celebrate progress over perfection; we decide to be proud of ourselves for how far we've come rather than bully ourselves for not having "arrived" just yet.

Our culture often focuses on the result rather than the process, but in the case of the unplugged hours (and most things in life), the really good stuff lives in the process. That's where the wonder emerges. That's where the joy takes shape. That's where the growth occurs. Yes, some moments will feel challenging and hard, but I encourage you to keep pressing into the process. Get curious about the feelings you're experiencing. Maybe even scribble down some notes to record what you're learning along the way. When all is said and done, it's you—learning to be present in the process as it unfolds—who has the power to really change the shape of your life. Wait and see. Just you wait and see.

1,000 UNPLUGGED HOURS

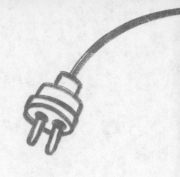

part one

POWERING DOWN

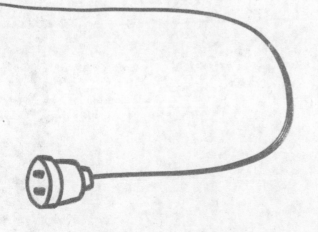

chapter one

STARTING POINTS

This story began on my thirty-third birthday.

I sat cross-legged on the carpeted floor of my friend's screened-in porch, breathing in the quiet stillness of the early morning. The humidity that sits heavily in the Florida air, always making sure its presence is felt, seeped into the open space all around me. The sounds of my husband's and daughter's laughter tangled together from inside the house.

I took a sip of my coffee, the steam rising from the mug, and quietly welcomed myself into another year of life. I've spent so many birthdays in this exact spot—surrounded by friends who've morphed into family over many years and seasons. This beach house has become a benchmark of sorts in my life. I've forged a tradition of sitting in this same spot, year after year, with a cup of coffee, a Bible, and a journal. I reflect on the previous year and build a fresh vision for the year ahead.

Most people simply enjoy a few well wishes on their birthday, eat some cake, and leave it at that. I envy those people, but I've never been one of them. I've always treated my birthday like a second New Year's—an excuse to concoct a dramatic plan to change everything about my life in 365 days or less. Since my birthday falls on the last day of May, it's almost the perfect halfway mark in the year.

Transformation has always felt alluring to me. I love the possibilities that come with change. Few things motivate me more than turning to the teeming potential of a blank calendar page at the start of a new month. But on this birthday, I was starting to wonder if the present moment—and who I was in it—would ever be enough. The previous year felt marked by a tiredness I couldn't quite name. I didn't have a vision for the year ahead. I felt exhaustion instead of excitement. Yet everything in my life was good. From the looks of things, I was doing just fine. But still, a growing weariness loomed beneath the surface, and I knew I would have to face it eventually.

I flipped between checking apps and responding to birthday texts as I attempted to set a goal for the coming year. Maybe I could grow my business. I could take on a new side hustle. Or I could finally become a runner. And that's when I felt it. A gentle nudge in my spirit seemed to come out of nowhere, impressing me with four words:

Turn off your phone.

That's the best way I can describe it to you—a nudge, a voice within that didn't sound anxious or tired. It felt strong yet somehow soft at the same time. It felt like peace and sureness—a sense that something good awaited me if I would simply listen to the instructions.

Turn off your phone.

I've lived enough years to know that these kinds of nudges—the ones that are almost audible—don't happen all that often. So when they do, I know to pay attention to them.

I turned off my phone at that moment, not even giving it a second thought.

I kept my phone off for the rest of that day. I wasn't anxiously checking it, wondering if enough people had remembered my birthday. I felt engaged in the day. I felt deeply connected to the people around me. I felt celebrated and seen, unhurried and at rest. For the first time in a while, my life felt like it was enough just as it was. I didn't need to fix it or change it or make it better somehow.

I experienced more joy in those twelve hours than I had felt all year—completely powered down and present. There was already something to these unplugged hours.

———

Something in me knew the unplugging wasn't a onetime thing. It wasn't a quick fix, as if I could turn off my phone for one day and reap all the benefits at once. It was more like, *Turn off your phone and keep turning it off. There are things to learn here. It's going to take some time.*

It was a call toward something deeper—an invitation away from the life of constant connectedness I'd built for myself and into something far better.

Over the days following my birthday, a challenge began to emerge. I settled on a goal of one thousand unplugged hours in one year. One thousand unplugged hours over one year equaled a few intentionally powered-down hours daily. I didn't want to be completely disconnected from technology, but I was curious to see if a balance could be fought for and eventually won. I defined what an unplugged hour would look like for me: no phone, no email, no TV or streaming.

I created a tracker composed of one thousand tiny bubbles. Each bubble represented a single hour unplugged. When you're forging a new habit or building a new lifestyle, it helps to document your progress, measure the process in some tangible way, and be specific with the results wherever you can.

I printed out the tracker and hung it on a set of antique school lockers in my office, where I would see it every day.

From there, I started turning off my phone one hour at a time.

———

The thing about the stories that change our lives is that they often have multiple starting points.

Sometimes, a story's starting point is clear and distinct. You remember what you were wearing or where you were standing or how life seemed to split in two in a single second. Other times, stories start at a dozen little junctures, and it takes a while for you to realize that the junctures were threading themselves together the whole time, trying to get your attention.

So this story may have started on that screened-in porch in the hot Florida air, or it may have begun over a decade ago during a media class in college. I can still remember my professor introducing the concept of "social media"—a significant cultural and technological phenomenon that would change everything about our world in the years to come. I thought, *Wow. This is going to be big. It will change so much about how we connect.* But I also felt an uneasiness stirring within me as I sensed that this kind of connectedness had the power to take so much from us if we weren't careful with it.

It's also possible this story began in a moment that happened soon before I felt that steady nudge telling me to turn off the phone. In one of my quiet times, I came across a line in Scripture that I'd encountered dozens of times before. In it, Jesus declares,

"I have come so that they may have life and have it in abundance" (John 10:10 CSB). Something about the phrase "in abundance" jumped out at me and I dug deeper. I wrote down several meanings: "exceedingly more," "over and above," "superadded," "something further," "extraordinary," "uncommon," "more remarkable." Then I wrote in my journal: "Is this the 'exceedingly more'? This rushed, can't-keep-up pace? This feeling like I always need to be 'on' and hustling toward something better?"

There was a disconnect between me and that abundance. I had a feeling it came from the hurried pace of technology advancing all around me and my thinking I had no choice but to somehow keep up—no matter the cost. The more I was faced with the hurry, the more I thought, *No, there has to be a better way forward than this. I have to find a way to get back to a place of peace and rest. I know that place exists.*

It might be that this story started years ago on an unremarkable evening in New York City as I walked up the stairs from the subway station at Bleecker Street. I can still remember dusk hanging in the air and Manhattan bustling around me with the intricate energy it exudes only during quitting time.

I checked my phone and found an email waiting for me, received sometime during my subway ride across town. Sent through a contact form on my website, the email was short— three, maybe four sentences. The writer introduced himself as Matt from Ohio. Just Ohio. Big ol' Ohio: population twelve million. That's the only detail I had about him—he left no email address for me to respond.

Matt told me how he was beginning to think he would leave no legacy behind. He wasn't sure his life meant much. But the last line of his email struck me, becoming a haunting thought I couldn't shake off. He wrote: *Everyone is so afraid of what will happen when the screen shuts off.*

I stood there by the subway stairs, people moving all around me at a restless, rapid pace—most of them with their eyes locked on their devices as bars of service reemerged. Matt from Ohio had somehow managed to put into words a fear many of us—myself included—didn't even know we had.

We've built so much of our lives around constant connectedness—around devices that affirm us, reward us, draw us in, and make us feel like our presence matters, even though, in an instant, they can also make us feel like we don't matter at all. Who would we even be without our screens, without this unlimited availability at all hours?

It's been over a decade since I got that four-line email from Matt from Ohio, yet hardly a day goes by that I don't think about that last line. I had no idea what would happen when the screen finally shut off, but I knew I had to follow the nudge and see for myself.

As it turns out, my whole life was waiting for me on the other side.

chapter two

MAPS

I've always been a lover of maps. I love how you can search them until you find that little dot or star, etched in ink, that tells you, "You are here." There's something quietly beautiful about knowing someone has stood where you're standing long enough to draw a picture of the surroundings and carefully give everything a name. It makes you feel like you can exhale, look around, and, when you are ready, find a way to move forward again.

So imagine these words you're reading right now are a map—unfolded, creased, and spread out on a table—showing where I was before the unplugged hours began.

The first place I'd point out to you is an area I've long called the "Growing Disconnect"—a landmass that spread, taking more and more territory over time. I felt the Growing Disconnect hovering in my spirit for years. It started as an inability to be fully present in the moment. With so many notifications coming through at all hours, there was always something keeping me from being fully present. Even though I knew my life was happening right in front of me, all the pinging and alerts convinced me that what was essential and urgent was happening on the screen. I fell into the habit of always needing to reach for my phone.

I'd document the moments. I'd share the moments on social media. But I never really felt like I was participating in the moments as much as I wanted to be. I was always somewhere else, consumed by thoughts of what other people were going to comment or like or engage with. If I'm being honest, the opinions and praise of other people shaped how I felt on a daily basis—how I showed up to my life off the screen. It was almost as if I didn't know how to assign value to my life experiences anymore—I was waiting for other people to do it for me.

It's no surprise the Growing Disconnect spread to my closest relationships. The more I connected online, the less energy I had to connect in person. I found it easier to stick to the surface in my relationships. It was less emotionally taxing to simply edit what I told people, giving them fragments rather than the full story.

I said things like, "I'm good! We're good! Busy! We're busy! Life is *so* busy!" Busy was my default answer through so many seasons of life. We live in a culture that praises "busy" even if we're on the brink of burnout, so it was easy to let that word spill out all the time. I look back and realize I used that word like a shield—it was something I could hide behind, a way to deflect attention rather than fully allowing people into my life.

When we edit the way we present ourselves well enough—with enough curation and filtering out of the honest parts—people don't ask the more challenging questions, like "Are you burning out?" or "Are you tired of trying so hard?" or, put very simply, "Are you okay?"

The Growing Disconnect snaked its way into family life. Our daughter, who was just a little over a year old when I started unplugging, had to have noticed I was picking up my phone every few minutes. It was always present, like an additional person in the room who could chime in at any second. Even though my daughter was so small, I imagine she might have wondered, *What's taking Mommy away from me so often? What on that shiny little object is so important to her?*

My husband and I weren't completely disconnected from each other, but we'd admittedly grown stagnant in our rhythms together. We'd put our daughter down for the night, finally cook dinner or order takeout, and sit down in our respective spaces—me on the big chair and him on the couch. We'd turn on yet another binge-worthy show. And before long, we'd look over at one another and find we were both scrolling. We were not fully paying attention; we were not chatting about the episode. We were existing, four feet apart, on our phones, and calling it a night together.

Many people I talk to say that's the trade-off for being so digitally connected—the feeling that no one is ever fully with us anymore. We're never fully present. We're on our devices at all times. We bring them to the table. We leave them beside us at dinner. We stop and start our conversations to check in and check out. My mom called me the other day to vent about a friend she'd met for lunch. Throughout the lunch, her friend was on her phone. Their whole conversation felt disconnected and impersonal. "Why did she bother going to lunch with me?" my mom asked. "It felt like I didn't even need to be there."

The double-edged truth about the devices we hold is that there will always be something to check. Something to read. Some way to improve. Something to watch. Another thing to reply to. Something to share. Another comment to make. I would often reach for my phone to do one thing and then find myself moving from app to app, completely forgetting why I picked it up in the first place.

Constant connectivity has become the norm, but I'm beginning to wonder, at what cost? Without our devices, we begin to feel isolated and anxious, disconnected from the rest of the world. There's even a name for this feeling—a condition called "nomophobia," the fear of being disconnected from one's mobile phone. The number of people identifying with this phobia is rapidly increasing worldwide.[1]

I think much of our desire for connectivity—especially via social media—is fueled by classic FOMO (fear of missing out). Years ago, I had a conversation with a friend who told me the one time he unplugged, his favorite artist played a secret concert at a venue right by his New York City apartment. Being unplugged meant he missed the tweet and the concert. He said he would never turn off his phone again—he would never miss another notification. I remember thinking that sounded like a miserable way to live.

What's wild is that research shows our levels of cortisol—the hormone that regulates our body's most important functions, including stress—rise even when a phone is merely in sight or when we think we hear one in the distance.

Clinical psychiatrist and founder of the Center for Internet and Technology Addiction David Greenfield explains, "The body's natural response is to want to check the phone to make the stress go away."[2] So we keep checking. Yet the things we check into often make us even more stressed. The spiral deepens.

———

The next sight your eyes might be drawn to on the map is the exhaustion. It was all over the place, like little interstates snaking in every direction. I found myself tired all the time. It wasn't a "new mom" tired or an "I somehow survived a never-ending pandemic" tired. It was a heavy soul-weariness seeping into everything: my relationships, my career, my faith, my view of life.

I think a lot of us feel this. I emailed my blog readers a few years back and asked a simple question: "Are you tired and worn out?" My inbox filled up with hundreds of messages from people all over the world. We were suddenly all admitting it, nodding and lifting hands to say, "Yes, I feel it. Everything feels like it's too much and it's too hard. I don't know how to navigate to a place of peace."

I believe much of our exhaustion stems from the information coming at us from every direction. We're living in an age of constant information overload. Before we even consume our morning coffee, we can connect with six people, debrief the state of the world's current affairs, answer half a dozen emails, get roped into yet another Zoom meeting, order a new sweater off Amazon, and bookmark a really great recipe for sourdough bread.

According to neuropsychologist Cynthia Kubu, our brains were designed for monotasking, not multitasking.[3] Never mind operating with twenty-three tabs open simultaneously. Slowly, without even realizing it, we lose our ability to focus as we permit more and more noise to enter in. Our brains were never wired to consume this much information at once. As a result, our attention spans get shorter, we're more likely to exhibit anger-prone behavior, and our mental health suffers.[4]

I often felt like my brain was in a fog I couldn't break through. Not knowing how to cope with the fogginess, I plugged in even more.

A series of pressure-filled thoughts popped up in my head the more plugged in I became, like splotchy bodies of water spreading out all over the map:

I need to be everything to everyone.
I have to hustle harder, or else I'll become irrelevant.
I can't take breaks, even though I'm exhausted.

Maybe you've encountered the same kind of noise. Perhaps you've been fueled by the anxiety that you are never doing enough. So you resist taking breaks and you barely rest. You press the pedal down harder and tell yourself to do more. You're convinced if you slow down then everything will stop. You've told yourself that it's all on you to hold everything together. *It's all on you.*

And maybe you cope with that churning anxiety the same way I did for a long time: Instead of getting to the root of the issues, you dig deeper into the trenches of technology, thinking it will make you feel better. More social media. More emails. More projects to take on.

I spent years placing impossible metrics on myself, believing bigger was always better and doing more was always the correct answer. It wasn't until I turned off my phone that I started asking myself better questions: Who are these metrics for? What would happen if I finally let go of them? Who would I become?

I felt so wrapped up in the frenetic pace of my life that a constant feeling of overwhelm seeped in. Other feelings bubbled to the surface too: comparison, jealousy, discontent. These feelings made sense, but the emergence of certain other feelings left me uneasy. I felt anger, but I had no idea where the anger came from. I also felt an overwhelming sense of apathy—I felt desensitized to things I once cared deeply about. The apathy was the feeling that scared me the most.

I've always had a big, bold, feel-all-the-feelings heart, and it felt like my sense of compassion was eroding from the steady onslaught of noise. I wanted to care about things. I desperately wanted to care, but I couldn't tap into my emotions. I felt debilitated by the constant stream of headlines and stories I was consuming online. Because I was feeling everything, I ended up feeling nothing. I felt distant from purpose, from passion, and from the faith I'd cultivated over a decade—the faith that had made me care so deeply in the first place.

———

Speaking of faith, picture a place on the map that is lush and green—a place where wild things grow. Imagine it as a wide-open space, like a clearing in the woods. But it's not so obvious on the map. It's tucked away and hidden, like a secret place that's a bit off the grid. That's how I pictured my faith for many years—as something beautiful, as a respite.

I've long believed that faith must be carefully cultivated with time and attention, but I was losing my store of both those things the more digitally connected I became. I was distracted. I was rushed and impatient. My time with God felt spotty and was easily interrupted by another notification or email.

In some ways, my faith started to feel like a performance. I knew my lines and could go through the motions, but the heartbeat underneath was growing fainter. The faith that used to hold so much mystery for me was becoming watered down as I tried to package it up for others to see. Getting to that place where our faith is more for other people than for ourselves can be easy in the age of social media. We feel like we need to comment or make a statement about everything. We're convinced we have to have all the answers, all the right words. And yet so much about faith is hard to wrap words around in the first place.

All the while, I found myself coming across the daily rhythms of spiritual giants—Thomas Merton, Therese of Lisieux,

Mother Teresa of Calcutta. Their faith seemed so real and raw. Connecting with God was a priority—the centerpiece that kept them stabilized—not an afterthought or something to get to if time opened up in the day. I kept thinking, *I want to get back to that place. I want to clear the space for better things—for hope, for passions, for exploration and mysteries, for things that actually matter to me—not this mindless scrolling.*

To tell you the truth, I initially had mixed feelings about sharing this map with you. The proactive parts of me wanted to skip right to the good parts of the story. But then I remembered that we often locate ourselves in the stories and maps of others. We gain strength from knowing someone else has experienced some of the same things we have. If our map makes just one person feel a little less alone in their surroundings, I think it is worth it to share.

Maybe that person is you. Maybe you found yourself nodding and saying, "Yes, that's me. I'm right here on the map. I feel the same things." If so, let me be the first to tell you the good truth that comes next: You're not stuck here. You're not lost. You can exhale, look around, and find a way to move forward again.

While maps show us where we are, they're also designed to spark a sense of adventurous hope in us—to make us feel that moving into new territory is possible. That's my hope for what's coming up ahead—that after some time has passed and we've both lived more of our stories, we'll come back to our maps and realize we've moved far from the spaces of emotional disconnect and exhaustion that once took so much from us. We'll have encountered better and more vibrant things along the way. Maybe we'll have drawn new maps entirely—ones that hold those better things all in one place. That, my friend, is the hope.

chapter three

TINY FORKS

I used to place so much emphasis on what I called the "big forks" in the road. Big forks are those significant life decisions everyone holds their breath for: where you'll go to school, who you'll marry, what job you'll take, where you'll end up raising a family, or how you'll choose to build your life.

Each of these choices is a fork—a splitting point between the life you choose and the life that doesn't end up being yours. But the older I get, the more I'm starting to see that much of our life's course is determined by the tiny forks in the road—the choices that are so small and subtle we're almost tempted to discount them. We face tiny forks in the road every day, and we think they're not that big of a deal, but enough of these forks—added together—end up defining who we become.

On the matter of tiny forks: There was a moment when things shifted for me.

My husband, Lane, and I were coming home from a date night. The whole evening had been relaxing and seamless, a much-needed pause for us to reconnect with each other in the thick of parenthood. But when we got into the car to drive home, a fight erupted. I don't know what it is, but it feels like there's something about date night, especially one that's going extremely well, that makes all parties vulnerable to a world war happening at the end of it. It's a phenomenon I don't understand.

I can guarantee with some degree of certainty that I picked the fight. There's a 99.8 percent chance that's the case. But I don't even remember what the fight was about—it was *that* petty and mundane. I just remember driving the winding back roads home in complete silence.

Lane and I hash things out differently. He needs at least a few hours to process before resolving things. People always say, "Don't go to bed angry," but I disagree. I think some people need a factory reset overnight.

I, on the other hand, the patron saint of ruining date night, want to fix things immediately, before they're even ready to be fixed. I'll throw out "sorry" in every direction to avoid the moments when we get quiet and don't speak to each other.

My attempts were of no use, though. The carnage of our perfect date night lay behind us. I reached into my bag to pull out my phone, to drown out the silence filling the car. I stared at the blank screen. I'd forgotten the phone was off. I'd intentionally powered it down at the start of the evening. Staring at the device in my hand, the truth struck me: I could stay in my momentary discomfort, or I could escape it with one press of a button.

I didn't know it yet, but variations of this tiny fork would keep coming up in the months and years ahead: each one a choice either to be fully present in my life or to escape it. I don't mean "escape" as though I needed an out from my life. My marriage was great. Motherhood was hard but not impossible. Work was good. It was just that over the previous few years, without even realizing it, I'd built a default mode of disconnect. I'd started tuning out at many junctures throughout the day, but I conveniently called it "connecting" or "working" or "self-care."

The reality is that when any discomfort emerged, I turned to my phone or busied myself with work instead of facing my discomfort. At first, these little escapes felt harmless. But as they started happening more, I began to see that I wasn't connecting; I was numbing myself. Author Brianna Wiest writes, "If you don't put the phone down feeling inspired or relaxed, you're probably trying to avoid some kind of discomfort within yourself—the very discomfort that just might be telling you that you need to change."[1] My ability to plug in became a portal I went through to avoid any emotion I didn't feel like facing: anger, sadness, jealousy, pain.

I don't think I'm alone in that. We bury our emotions in emails, texts, memes, and videos. If there's a feeling we'd prefer not to face, there's a way—right at our fingertips—to push away the distress and scroll mindlessly. The devices we hold make it so we never have to slow down, be alone with ourselves, or find a way to sit in the uncomfortable stillness that is supposed to come with daily life.

When you begin pressing into your own unplugged hours, don't be alarmed when discomfort rises in your spirit or you suddenly feel like you don't know what to do with your hands. We're so used to always occupying ourselves that we've all but forgotten what it feels like to just *be*—without a task, without a comment, without a point of engagement.

As that weird feeling of "What should I do with myself?" begins to envelop you, I'd encourage you to sit in it. Embody it. Let it wash over you and help you realize it's okay to just exist for a moment or two without an agenda. Being present in the moment is enough. And then I'd share with you the trusty words my college mentor once gave me: "Get comfortable with feeling uncomfortable." That advice rings even truer over time.

Resistance rising within us isn't a sign that we're doing something wrong. Often it means we're doing something right. We're on course. We met a tiny fork, and we chose the road that would grow us beyond who we'd been up to that point in our story. Our uncomfortable feelings—the ones we most want to escape—often yield the most significant growth moments. But there's a catch: We must stay in the discomfort long enough to see hard-earned growth emerge.

Back in the silent car after our date-night fight, I glanced at the blank screen in the palm of my hand one last time. Press in to my discomfort or escape. Those were my options.

I placed the phone back in my bag. The moment was so small—a tiny fork. No one but me even knew a choice had been made. I settled into the silence, into the discomfort, and it's strange to admit, but I felt only peace for the rest of that car ride. That was the start of pressing in.

chapter four

CHECKING IN

———————

In my writing classes, I talk to my students about the importance of getting into the writing room and shutting the door—
closing out the outside world so they can get alone, explore the
territory of their minds, and see what's waiting there. It's the only
way I know to experience a breakthrough—get into the writing
room and shut the rest of the noise out.

A passionate follower of true crime since I was ten years old—
prone to pinning up autopsy reports on my childhood bedroom
walls beside glossy centerfolds of Hanson—I use the analogy of
a crime scene to further my point. When a crime occurs, one of
the first things officials on the scene do is rope off the crime scene
with bright yellow caution tape. They do this to establish the
perimeter of the crime scene and to secure evidence that might
be present. Investigators are looking to protect the integrity of
the crime scene by monitoring what goes in and what comes out.

I tell my students we must do the same thing with our lives.
We all have an inner world, and every day is a battle to protect

the integrity of that space while operating amid so many exterior distractions. We must become ruthless gatekeepers of what goes into and what comes out of that inner world.

———

At the beginning of my unplugged hours, I felt like I was spinning in a cycle of constant overwhelm. I felt anxious and stressed most of the time. Much of the overwhelm came from the pressure I was putting on myself to be everything to everyone, all at once. At first, I took this quest to be "all the things" as a challenge. I could be the mom. I could be the CEO. I could be the good wife, the intentional friend, and the kind neighbor. I could write, and I could cook. I could change hats constantly.

I'd often get comments from people marveling at how much I managed to accomplish. Those comments kept me moving. I didn't care that I felt ready to snap inside, so long as no one could tell from where they were standing.

During that time, I'd wake up and immediately check my phone before doing anything else. Before I could properly open my eyes or even turn to look at my husband sleeping beside me, I was scrolling. I was allowing other people's fingerprints—their agendas, opinions, praise, and problems—to get all over my day before my feet even touched the ground. It's no wonder I woke up with anxiety that settled like a thick dew over every task I set my hand to.

I would fall asleep with my phone next to my head and wake up to check on things, sometimes even in the middle of the night: new emails, comments, likes, notifications. And while there was nothing inherently wrong with being on my phone, the things I focused my energy on would inevitably set the tone for my day.

I moved from "checking things" to consuming other people's

content without even noticing it. I thought I was checking in with the world around me. I look back now and realize I was checking out.

And it wasn't just social media or emails. I'd check retirement funds. I downloaded a game where you earn virtual money to build your virtual house through playing virtual games, and I'd waste hours perfecting my digital mansion. I'd check couponing apps and weather apps. At one point, I convinced myself I would become an investor and started dabbling in the stock market. I had no clue what I was doing, but it gave me an excuse to always check my stocks—anything to avoid checking in with myself.

It's not our fault that we check in and out of things so frequently. Our brains are wired for dopamine—the chemical that allows us to feel pleasure. Daily life brings all sorts of natural dopamine releases—exercise, sunlight, conversing with a friend. But our devices deliver intense doses in less time than it takes to lace up our shoes or get outside. So we scroll. And we click. And we shop. And we search. And our brains get hungrier for something more. But we keep reaching for the easiest surge—getting cheap dopamine hits from a device that has programmed us to check out of real life more and more every day.

———

In Psalm 42, believed to have been written by the sons of Korah—a family of Levites who were known for their temple worship—the author cries out what has become one of the most well-known refrains in the Bible: "Why are you downcast, O my soul?" (BSB). We quote this line so often—slipping it into sermons and devotionals—that we sometimes water down the mystery and the miraculous waiting for us within these words.

The author isn't crying out to God—not yet. Later in the psalm, he will fix his gaze back on God, but first, the author speaks to his own soul. He pauses, turns his attention inward, and checks in. He takes the time to ask himself some critical questions:

> *What's really going on?*
> *Why are you overwhelmed?*
> *Why are you spinning your wheels so hard?*

What I find beautiful is that Psalm 42 is considered an instructional song. It was written to teach us how to navigate difficult passages in our waking, breathing life.

We are meant to check in with ourselves. We are meant to ask ourselves tough questions. We are meant to look inside, survey the chaos, and ask ourselves, *Are you okay? What's going on? How can we fix this?* We're meant to know what's going on in our souls, to dredge up our old storylines from the murky waters and face what's really there.

I feel like we talk a lot about checking in with others but hardly ever about checking in with ourselves. What would that even look like? And doesn't it sound a little selfish? I mean, checking in with ourselves isn't exactly productive, and on the surface it may seem like there are much better uses for our time. There are kids to feed. There are emails to shoot off. There are deadlines to meet and goals to achieve.

And I'll be honest: The last thing I want to do is check in with myself if I don't want to admit something isn't working. No thanks, I'd rather drown myself in memes and watch endless videos of surprise soldier homecomings.

The checking in began slowly for me. I bought an alarm clock. I placed my phone in another room before I went to sleep at night. I got up earlier. I stopped waking up to the droning scroll and changed the order of what I consumed first thing in the morning—from emails, comments, and text messages to coffee, Scripture, and sunlight. I accepted that I'd forgotten how to sit and do nothing at all. How to wait in silence. How to be—just be—without telling anyone where I was with a picture or a post.

Slowly, I'm regaining this ability—and feeling more alive.

For a long time, I thought being alone was the same as loneliness, and I did everything in my power to avoid it. We confuse the two all the time. Being alone is an ability many of us lack in today's noisy world, but solitude is where we often uncover the most strength for our days, for the challenges we face, for the dreams within us to take shape. In his book *Together: The Healing Power of Human Connection in a Sometimes Lonely World*, nineteenth and twenty-first surgeon general of the United States Vivek H. Murthy, MD, explains how our spirits need solitude. We need to be able to replenish and restore and "connect with ourselves without distraction or disturbance."[1]

I'm learning that checking in with our lives isn't about getting away from all the noise entirely. It can't be. There will still be plenty of days when distractions come from every angle, and we need to find a way to root and ground ourselves in the present moment. To turn inward, even amid crying babies, dirty dishes, and unread emails. We are worth taking the time to do a little pulse check on ourselves.

For that reason, my checking in has started happening in smaller spurts and through unconventional methods. Sometimes, checking in looks like drinking a big glass of water or remembering to eat something that contains a nutrient or two. It might look like scribbling down a quote I'd like to read again or listening to a voice memo from a friend. Right now, checking in for me looks like walking to the end of the street and back between writing sentences like this one. It's enough time to feel the sun's warmth on my face, to recalibrate and get the energy I need to step into the next part of the day. The next sentence. The next email. Whatever the next thing might be. At all these little junctures throughout the day, I'm checking in with myself when it would be easier to pick up my phone and check out.

The other day, I intended to do a great big check-in. I had the space for it, and I felt it coming. I placed my journal and Bible on the kitchen counter. I made a cup of tea. I got the chance to stand at that counter for approximately three sips and three verses. And then, well, I'll spare you the details, but you can insert "explosive bowel movement," "tiny toddler," and "floor" into the Mad Lib.

I ran a warm bath for her. I stepped back into the clinking, clanking chaos of daily life. And maybe it didn't fill my soul to the point of overflow as I'd hoped, but I could see the stray cup of tea and the laid-open Bible on the counter for the next few hours, and it filled me with a strange sense of joy to know I'd carved out this space for myself.

———

What would happen if we checked into the lives we're building as much as we checked into other people's lives online?

What would change if we stopped idolizing the lives we witness on digital platforms wired for curation and perfection and stepped into our own unedited lives?

How much could we flourish and make happen if we stopped waiting to feel fully sure of ourselves and fully ready, and just stepped into our lives exactly as they are?

These are the better questions I'm asking myself lately, and I am already starting to see the answers to them: *So much.* That's what would happen if we checked into our own lives. So much would flourish. So much would finally take shape. So many of the worries, doubts, and fears might fall away if we began checking in the same way we've grown used to checking out.

We may learn some things about ourselves. We may find out we're bone-tired. We may figure out we feel forgotten and despairing. We may discover that we're craving something softer, slower, or more rooted. We may discover that we're done trying to keep up in a race that never came with the promise of a finish line.

Author Anne Lamott writes, "Almost everything will work again if you unplug it for a few minutes, including you."[2]

So breathe. Power down. Check in.

You just might be trying to get your own attention.

The noise isn't going anywhere—I can promise you that. It will all be here when you get back. But maybe you'll be the one who has changed.

chapter five

EXACTLY WHERE
YOU NEED TO BE

As I write these words, there's an art print hanging on a bulletin board directly in my line of sight. The words "You are exactly where you need to be" are etched in swooping golden cursive on black cardstock.

This art print was the first thing to enter our empty house when we moved in. Our best friends showed up the moment we got the keys, and they walked the barren hallways with us for the first time. My friend Dawn handed me the art print with a note and the date written on the back to welcome us home. We popped a bottle of champagne together and slurped the bubbles out of tiny plastic cups—a toast to new beginnings.

This art print is mobile now—it gets placed all over the house at different times of the year. I move it based on where I need to see the words the most, or where I think someone else needs to see them.

At this point, it's a part of the house—as essential as a beam, light switch, or floorboard. Of all the messages I've needed to rinse and repeat over the years, this is the one that needs the most continual practice.

You are exactly where you need to be.

It's so easy to doubt.

We operate in a world where there's always something seemingly better just beyond our fingertips. There's a new gadget. A trendy wellness practice. A new app. A better diet. A new car. A different home. As our social media algorithms work overtime in the background, they learn to read and know us in order to cater to our tastes and desires. They start advertising directly to us whenever we search for something. We have new options at our fingertips at all times.

With all these potential options surrounding us, it's easy to believe that our lives—the ones we're *actually* supposed to live— are lost somewhere in the future. Not these hard, crusty lives where pain is inevitable, prayers feel unanswered, families hold drama, marriage is complex, and the dream job isn't so dreamy. Indeed, our best lives are waiting for us elsewhere.

——————

For years, I thought my life was on pause until _____ finally happened. The blank is different for all of us. Life will start happening when:

the job comes through.
the right person walks in.
the bassinet is filled.
the jeans fit.
the healing comes.
the ring is on the left finger.

So long as you think the answer is in that thing, I'm afraid it will never be enough. When you manage to get that thing, a new something rises to fill its place. Another hole will always need filling until you decide to stop digging holes altogether.

———

Earlier this year, I attended a conference where one of the speakers talked about contentment. She pointed to some famous words from the letter Paul wrote to the church at Philippi: "I have learned the secret of being content" (Philippians 4:12). She drew out the word *learned* from the passage. She said that's the thing we never talk about—the learning curve.

We often talk about contentment as if it's already wired within us. Like it's just waiting for us to wise up and flip the switch. If that's the case, I'm missing parts. But Paul doesn't claim he became content with the snap of a finger. He *learned* it, meaning he practiced it like anything else. There were likely stops, stalls, do-overs, and grace to cover it all.

Learning to be content with our circumstances—what we have and who we are becoming—is an ongoing process, not a onetime occurrence. It's a dance we must engage in every single day. And even when it feels like we've learned the steps by heart, our life might introduce a new move to the dance in some later season, and we'll have to learn that one too.

During a recent coffee shop visit, I met a young woman who had deleted all forms of social media for nearly nine months. I found this out because I dared to strike up a conversation with her—a habit I've been awkwardly trying to cultivate since the pandemic. I think that time of isolation led many of us to realize just how much we craved one another's presence—how much we'd taken it for granted in the past.

Before long, she and I were sitting directly across from each other, legs crossed like children during circle time and cups of coffee in our hands as she told me her story. When I asked her what breaking point had finally made her delete the applications, she told me that scrolling and looking into the lives of others had started to tell her a story that her life—one that had seemed perfectly good before—was no longer enough.

Soon, that feeling began to spread to other parts of her life. She started to feel like maybe there was someone better out there for her, someone other than the loving, amazing spouse she already had. She saw so many couples projecting their lives online for others to see, and she wondered if she was meant to be with someone who acted more like the people who were entertaining her online.

That was the breaking point, she said. When the discontent spread from just her and latched tightly onto someone she loved, someone who had always been her first choice in the past, she knew something had to give. And so, she decided to step away from the noise to preserve the beauty of what she already had. When she did, she rediscovered the truth that had been there the whole time: Her life didn't need to look like someone else's life on-screen, but her life did need her to pay attention to it to flourish properly.

———

I learned the other day that when monks enter the monastery, they take four vows. A vow of chastity. A vow of obedience. A vow of poverty. But not as widely known and completely new to me is the fourth vow: a vow of stability. With this vow, the monk roots himself where he is, promising that he will not move to another monastery unless he is sent there by a superior.

Thomas Merton, an American Trappist monk and great theologian, wrote,

> By making a vow of stability, the monk renounces the vain hope of wandering off to find a 'perfect monastery.' This implies a deep act of faith: recognizing that it does not matter where we are or whom we live with. . . . All monasteries are more or less ordinary. The monastic life is by its very nature 'ordinary.' Its ordinariness is one of its greatest blessings.[1]

There's that dance of contentment all over again.

Something about the vow of stability draws me in. I find myself stopping at lulls in the day to turn it over in my head and ask: What does that look like in my life? It often feels like it would be easier to be somewhere else. But I think there is value and weight to committing to the life we actually have when it feels easier to escape it by distracting ourselves with another app, another influencer, another show on the newest streaming platform.

There is power in looking at the life before us—exactly as it is right now—and deciding to embody it fully rather than spend another moment believing that something better might be waiting around the corner. Another option. Another more beautiful thing. Another patch of greener grace. We must decide to embody the good, the bad, the holy, and the hard long enough for our roots to grow deep into the ground and strengthen us into stable forces in an often-unsteady world.

My favorite part of the day has quickly become the sliver of time between late afternoon and early evening when I close my computer. Before the unplugged hours took shape in my life, I moved into my evenings seamlessly but still completely connected. I checked in and out throughout the night. Somehow my work always managed to come with me. I'd check emails into the late-night hours. I could always find another task to complete. Now I'm practicing uninterrupted time—a pocket of two to three hours each night when I'm away from my computer and my phone is out of sight.

I thought the unplugging might make me anxious, but a hard-to-describe-with-words feeling has been meeting me when I swipe my finger across the screen to turn my phone completely off. It's brief but feels like a giant exhale, a release of something I've been holding on to all day.

It's a feeling of rootedness. It's not the feeling that life is perfect or that problems don't exist. It's not a feeling that stress will never return or that emails won't come in overnight. Instead, it's a feeling of deep appreciation, one I've missed for years, that starts to spread over me. It's an appreciation for what exists right now. It's a feeling that I don't have to keep digging for something better; I can put down the shovel and learn to love what's right here.

My mind quiets, and with no more noise coming in, a knowing inner voice pipes up and quietly says, "It's all right here. Look around—it's all right here." Suddenly, there's space. There's time. There's potential that I overlooked before. There's a stability I've been searching long and hard for.

And so, while I'm not an expert on contentment, I think it starts with learning to take breaks from all the noise we've let in over the years. The noise that told us our life was waiting for us somewhere else. The noise that told us we had to be constantly looking for the next thing or doing yet another thing to hold on to our value and worth. The noise that told us we were missing vital pieces of ourselves—that we had to keep searching for ways to be better.

When you break from the noise, you start to see what's underneath it—the pieces of your life that have been right there all along. This is the raw material you're working with. Now the work begins.

chapter six

A PRAYER
FOR TEARING
MUSHROOMS

I t's a Wednesday morning, and I'm prepping vegetables for an evening meal.

Nothing can sway my love for a good slow-cooker meal, especially when all the recipe calls for is to take your ingredients, dump them in the pot, and turn that baby on low for the next eight hours.

That's what I thought I was getting myself into when I pulled the cookbook off the shelf. But alas, I am staring at a recipe that doesn't ask me to chop the mushrooms. No, no, it wants me to *tear* them. With my hands. One by one. Rest assured, there is no efficient way of doing this.

I am just about to give up and grab the knife when I remember something Lane told me about how the specific methods within the recipe matter, how you shouldn't mince the garlic when the recipe tells you to smash the garlic. Something different is released when the garlic is smashed.

In an act of defiance, I had looked this up to try to prove my point, but Lane is right. *Bon Appétit* recommends using your hands rather than knives whenever you can. When you gently tear a food, its texture changes and it cooks differently than when you chop it.

Alas, *Bon Appétit* neglected to report that tearing mushrooms is a completely arduous process that will make you want to stab something with a fork.

I've been at this mushroom-tearing part of the recipe for about five minutes, which is long enough to know I don't have time to tear the mushrooms. Who does?

———

Since the start of the unplugged hours, I've tried to embrace monotasking—being fully present in the one thing rather than trying to tackle a slew of things all at once. Clearly, I haven't made it that far in my transformation.

The thing is, I've been multitasking for as long as I can remember, and it hasn't made life easier or better. I think it's made me even more overwhelmed. Getting more done doesn't mean I eventually reach the end of the task list. It just means the task list gets longer, and I discover more things I "should" be doing—which explains the near-constant ball of anxiety sitting in my chest these days. I can never do enough. There is always more to be done. The list only gets longer. The ride gets faster.

Catherine Doherty, a social activist and spiritual writer, would say that the only antidote for life in this fast-paced, anxiety-driven world is to focus on performing the duty of the moment. She defines this duty simply as "what you should be doing at any given time, in whatever place God has put you."[1] The duty of the moment is always shifting.

The duty of the moment is tearing the mushrooms.

In a few minutes, it will be an email.

In an hour, it will be a project for work.

In three hours, it will be the carpool line.

So often, the duty of the moment isn't notable, striking, or worth documenting.

Paying attention to a single, unassuming task is its own act of love. It requires dedication, and that's the essence of devotion. There might be a voice in the back of your brain that whispers you should be doing more, or you should be juggling three things at once, but that's often just the voice of fear trying to get you away from the task at hand—from where you need to learn to practice presence.

———

These days, I keep my copy of *The Practice of the Presence of God* by Brother Lawrence always somewhere within reach. The book is falling apart. The pages are yellowing and coming undone. At this point, it's one big underlined and highlighted mess, but it's my favorite book and my best traveling companion for this season of unplugging.

Just under one hundred pages, the book can be read in a single sitting, but be warned, you will likely find yourself turning the material over in your mind for days and weeks to come. Brother Lawrence was a seventeenth-century Carmelite monk and friar known for his profound devotion to God through simple daily tasks. After he died in 1691, his collected letters and conversations were published in this small book. I call it a manual for how to find God in every moment. In the good. In the hard.

In the overtly sacred and the entirely ordinary—and especially in the ordinary.

Brother Lawrence gave up prayer strategies and techniques (except for intercessory prayer) and learned to find the presence of God in daily life. As he scrubbed dishes, he learned to communicate with God and give him his full attention. It was there—in the humdrum—that he encountered the sacred. He wrote in one of his letters, "I keep my attention on God in a simple, loving way."[2]

His experience is a refreshing reminder that I don't need to travel away from the moment or task at hand to find God. Habits. Silence. Daily rhythms. Yes, God meets us in these things. But he's not confined to a morning quiet time or the pews of a church. He meets us in our exhaustion. He meets us in our semi-sleepless nights. He meets us in our unfinished stories. He meets us in seasons of disappointment and in spaces of doubt. His presence is continuous, waiting for us in the laundry piles, in the making of lunch, in the early-morning workout, in the commute home, in the decision we have to make tomorrow.

The more I unplug, the more I believe that the idea of "praying unceasingly" isn't about keeping a perfect posture of prayer at all hours of the day. It's about repeatedly inviting God into the mundane moments of the day.

Crying out one word—*help, why,* or *please*—over and over again until you fall asleep—a prayer.

Staying up through the night, wrestling with a depression that feels endless—a prayer.

Facing crippling self-doubt but deciding to sit down and write for the first time in years—a prayer.

Feeling tired and strung out from weeks of not sleeping, doing everything you can to figure out this whole "motherhood" thing—a prayer.

Sitting with a friend who just miscarried for a second time and having no idea what to say to her—a prayer.

Tearing mushrooms for a meal that will nourish you after a long day—a prayer.

———

My shoulders relax into the present moment. It's just me and this mountain of mushrooms. No podcast is playing in the background. No reality TV show is propped up for me to watch as I cook. My full attention is here. A prayer begins.

> *Dear God, thank you for this meal I am preparing. Thank you for the family I am feeding tonight. Thank you for being in every task I engage in today. I might not acknowledge you, but you're there.*
>
> *You're in the coffee cups and the vacuuming of the hardwood floors. You're in the load of laundry and the invoices to be filed.*
>
> *Teach me how to be more devoted and present in the day-to-day activities that I'm tempted to discount.*

I sink deeper. The peace spreads. My mind drifts back to the times I prayed the rosary in college. I wasn't Catholic, and my beliefs about God were questionable at best, but I would go on Catholic retreats because the devotion of the people there fascinated me. I didn't know how to love God, but I wanted to be close to people who did.

Each night of those retreats, we'd sit together in one room and pray the rosary. Though I didn't know the inner workings of the practice, I loved the challenge of sitting still and focusing on just one thing. One bead at a time.

Each bead was a reminder that not every second will feel holy—there will be plenty of moments when my mind wanders all over the place—but there's an invitation to keep returning to the focal point when my mind has wandered off the beaten path.

I come back to the heart of things one bead at a time. One mushroom at a time.

Teach me to find you, God.

   In the making of a meal.

   In the telling of a bedtime story.

   As I'm wiping down the countertops.

   While I'm helping a neighbor.

   Teach me to see you in the mundane tasks when I'm tempted to say, "I don't have time for this." I want to do these things well.

   Please give me eyes to see you operating in the routine and repetitive moments of my life. Indeed, you are in it all. And thank you for the mushrooms—for the chance to get off the merry-go-round of multitasking and return to the duty of the moment. Help me to find value even here.

chapter seven

AN ODE TO MAGGIE

During my senior year of college, the campus club admin-
istrators asked all student leaders to participate in a day
of service. We scattered ourselves across the city of Worcester,
Massachusetts, sweeping sidewalks and painting the walls of
community centers.

A few of us got dropped off at a senior citizen center to visit
with the residents for the morning. Off to the side of the lobby
stood a table where a group of older women was playing cards.
One of the women motioned for my friend and me to come over
and join them. They scooted their chairs closer to one another
to make room.

"The game is gin rummy," the woman at the head of the
table announced as we pulled up two chairs. "I'm Maggie." She
introduced the other women around the table while dealing out
the cards.

With a full face of makeup and her hair done in curls, Maggie
was the last person I expected to encounter in a senior citizen

center. She talked loudly and with authority, voicing her opinion even when it didn't seem like anyone was asking for it. I didn't have to sit there very long to realize Maggie was the queen bee at this table.

During one of Maggie's turns, my phone buzzed on the table beside me. I quickly placed my cards in my lap and snuck the phone beneath the table to check my messages. I sent a quick text back.

"I don't understand all you young people," Maggie said brashly. I looked up to find her staring right at me, the comment lobbed in my direction without a hint of hesitation.

"You're always talking to one another on a screen, and I don't understand it. My granddaughter talks to her best friend on a screen. That is not a best friend!" She raised her arms, her cards still held neatly in her right hand. "You need to be able to see your best friend, touch your best friend, *smellll* your best friend." She moved her body in a circular motion like she was getting a whiff of everyone around the table.

As a storyteller, I want to write a version of this scene between Maggie and me that plays out differently from what really happened. I could say I savored her wisdom, and we talked about technology and getting back to the roots of human connection. But I was twenty-one and at a point in my life when I was pretty sure I knew everything. I didn't think much of her comments at the time, and I didn't care enough to realize Maggie wasn't just spewing her frustrations over the younger generation; she was sounding a wake-up call for what was to come.

This was 2009—the year experts say the smartphone caught on like wildfire. "Smartphone rises fast from gadget to necessity," the *New York Times* reported.[1] At that point, social media was merely a way to connect with people who were already in your day-to-day life.

We'd yet to see an era when celebrities could be born overnight through a viral video. An era when we could talk, shop, exercise, work, and make art via a small computer tucked in our back pockets. An era when we'd forget how to wonder because Google has an instant answer for our every question. An era when relationships can start and end on a screen without either person ever knowing the sound of the other's laughter.

Admittedly, I lost my way in all of it. The shift toward dependency on screens was gradual, but I eventually started believing I could cultivate my best relationships online. I could offer people a curated version of myself and think I was still being honest. I could go to bed with my phone in my hand and not call myself deeply lonely at the end of the day.

According to the Cigna US Loneliness Index, 61 percent of adults report that they sometimes or always feel lonely. Men are lonelier than women. Younger people are lonelier than older people. Those on social media have more robust feelings of loneliness.[2]

I was part of that 61 percent when I began turning off my phone. I might have been the leader of the 61 percent club, or at least the person who would have fundraised for the club jackets.

When the COVID stay-at-home restrictions lifted, there were parts of me that didn't want to return to normal. I did not want to leave the house. I didn't crave social situations anymore. I preferred to orchestrate my relationships from a screen. Maybe I thought I was getting what I needed from the constant pinging of my iPhone.

I think relationships naturally feel easier from behind a screen. I could offer little touches to many different people in a day without putting in real effort. I was casting a wide net instead of a deep one and didn't realize how much of a toll it was taking on me. How fragile my relationships were becoming. How much I was isolating myself but claiming I was surrounded. I had thirty-seven text threads per day to prove I was surrounded.

As someone who deals with depression, I know this isn't the answer. I know that isolation is never the answer. As much as my phone can feel like a meaningful form of connectivity, I still have to push myself into face-to-face human interactions to be the healthiest version of myself. And although I wish I could tell you I found a magical switch to solve this problem, I didn't. I had to start forcing myself back into social situations with a "What Would Maggie Do?" mentality.

It's been more than fifteen years since I met Maggie. Our interaction was brief, at best, but some people's wisdom has that staying power. Some conversations travel with us through life, sticking close like a second skin, returning to our memory occasionally to nudge us awake and make us pay attention. Maggie does that for me, even now.

Maggie tells me to get off the phone.

Maggie coaxes me to make the in-person coffee date.

Maggie reminds me to make lists on my phone containing the small details of the people I love—what makes them tick, what annoys them, what lights them up inside. All of it.

Maggie pushes me out from behind my screen until I remember that I am made for other people. Made to see them. Made to hold them. Made to sit with them.

Maggie pushes me to ask the hard questions:

What if I'm missing something critical?
What if the connections I need are happening all the time, all around me, and I'm just missing them?

What if those brief yet necessary connections are ten times more sacred than whatever connections I'm constructing online?

I may have missed out on the deeper conversation I could have had with Maggie, but I won't miss other connections that are meant for me.

———

Shortly after I started unplugging, Lane suggested we get out for a spontaneous date night. Since having a baby, my spontaneity has severely fizzled. Spontaneous, to me, is getting takeout from a sushi restaurant we haven't ordered from before and maybe, just maybe, getting a little reckless by switching up our Netflix selection for the evening.

We picked a reliable spot—one of our favorite restaurants, one that plants firepits all over the back patio during the colder seasons. Come November, they start projecting old-time holiday movies on the brick walls of the building.

We found a firepit toward the back of the patio and began our pre-dinner date night ritual: a few glorious minutes of sitting and reading in silence without anyone asking us for a snack.

Not even five minutes after we cracked open our books, an older woman with a bold pixie cut and wire-rimmed glasses strolled up to our firepit. A regal-looking greyhound stood on a leash beside her.

"Do you mind if I sit here?" she asked, motioning toward the Adirondack chair across from us. "I promise not to say a word."

I looked around the patio—plenty of open spaces and vacant firepits to pick from. We invited her to take a seat. She ordered a glass of wine, and we ordered an appetizer. We returned to our books, reading approximately two sentences before the woman started a conversation.

"Are you from around here?"

At that moment, I desperately wanted to just keep reading. To shut out the small talk with the stranger and go about our date night. But the little Maggie on my shoulder made me close the book.

I sat up straighter. And I listened to the voice inside me that whispered, *Just pay attention.*

The woman's name was Leah. She lived in the neighborhood, directly across the street from where our daughter had just started preschool a week earlier. She was from Florida. She had two daughters who were my age. She was a proud "granny" to two identical grandkids with fiery red hair. She pulled out her phone to show me pictures. From that point forward, she referred to herself as the "substitute granny"—making it clear that she was happy to step in as a babysitter whenever we needed her.

The more she spoke, the more I wondered about her. I wondered if she was lonely. Maybe she came to this spot often, not to sit silently by the fire but to try to find people to talk to.

Our conversation moved from mental health and loneliness during the pandemic to God, motherhood, and careers. She told me how she always struggled to balance motherhood and her career—a balance I'd been trying to find myself—but she felt she needed both to feel like her fullest self. And so she chose them both.

I gulped back tears at certain moments and cried openly at other points in the conversation, feeling refreshed by her presence. The regal greyhound sat stoically beside her the entire time.

At times, I felt like I was sitting with an older, future version of myself—someone who'd lived a lot more life and had many more scars but was telling me I would be okay. Assuring me that life wouldn't be void of hardships or seasons that felt unsurvivable, but that I would emerge stronger, tougher, and more resilient than I ever thought possible.

Our conversation lasted two hours. My phone stayed face down the whole time. We exchanged numbers before parting ways. My phone pinged almost immediately after she walked away.

"I'm the substitute granny!" she wrote. Smiley face emoji.

As we walked away from the firepits, I couldn't tell if those two hours were more for her or for me. It didn't matter. After years of feeling like spontaneous, in-person connections would never happen again, this was a gift. A divine meeting I easily could have missed.

Lane and I found ourselves back at the same restaurant a few months later, with our books in hand. We sat at a small table outside the restaurant. A few tables down, we noticed a couple sipping on cocktails and talking with a familiar woman and her greyhound. It looked like she'd been sitting at their table with them, but she walked away a few minutes later.

I asked the couple if they knew her. The man shrugged his shoulders and took another sip of his drink. "Never met her before," he said. "She just asked if she could pull up a chair."

I smiled. The substitute granny was at it again.

I'm convinced the digital route will always be the easier one. It will always feel like it's costing us less to stay close to our devices and forge relationships that come with filters and backspace buttons. But then I remember Maggie, and people like the substitute granny, and I know that the digital route is missing something.

Whether we're willing to admit it or not, we desperately need the physical presence of others. We need to be able to sit across from one another at tables, meet up in coffee shops, and find our way toward one another at random firepits. It's that old Ram Dass quote put into practice: "We're all just walking each other home." And while there are apps for nearly everything in life—from apps that tell you the weather to apps that report the best time for you to run to the bathroom during a movie—there is still no app that can simulate the human connection we need to keep us moving forward. The way it feels to apologize in person. The thrill of meeting someone for the first time, thinking this person could be someone different from all the ones you've met before—your palms sweating and heart thudding loud in your chest as you see flickers of a future with them. The incredible beauty of fumbling through hard conversations without any backspace or delete buttons, fighting to be known through the awkward pauses and the elephants in the room. These are the things that make us fully human, and they fuel us to keep going in a world where it's easier to give one another the two-dimensional versions of ourselves. Maggie would tell us the 3D version is better.

I hope Maggie is still out there, ruling the roost at the senior citizen center. If she is, I want to tell her, "I'm out here too, Maggie. I took your advice." If I get to be her age—even if we're at the point in technology where holographic versions of my grandchildren come to visit me at the senior citizen center—I want to have a thick collection of moments to show for myself, moments in which I was fully present and connected. I know I have to fight for those moments, but I want them for me, and I want them for you too.

Moments where we pay attention to the people unexpectedly placed in our pathways.

Moments where we make small talk with the strangers in the checkout line because we need those tiny, brief connections just as much as we need deeper relationships. And sometimes other people need them even more.

Moments that are so good and rich and connective that they make us forget to check our phones entirely.

chapter eight

I WAS HERE

I spent much of my early twenties consumed by what is now referred to by researchers as "purpose anxiety."[1] It's good to have a name for something I always just called a quarter-life crisis. The term encapsulates the distress that people, both young and old, feel as a result of wanting their lives to mean something. With the rise of social media, I imagine purpose anxiety has only ramped up. It's incredibly hard to maintain a solid sense of purpose when you're rapidly consuming what everyone else is doing—how they're succeeding and how they're moving forward.

Back then, I believed purpose had to be big and noticeable. I thought my life had to be flashy and shareable. This often left me feeling like I was waiting for my life to begin—waiting for that big moment to unfold when I would finally "arrive."

When I lived in the Bronx after college, I spent gaps in my workdays down the street from our apartment at a little library. The library only had a few shelves, but it had one specific book that was the reason I kept going back. It was a reference book, which meant I couldn't borrow it, but I would pull it from the shelf and sit off in the corner flipping through the pages. The book was called *Portraits*, and it held a series of 1,901 profiles published in the *New York Times* after September 11, 2001. I'd trace the intimate, beautiful details of these individuals whose lives had been cut devastatingly short. Profile after profile, what struck me was that the family and friends rarely, if ever, talked about the big moments or even notable successes of their loved ones. They spoke of the smaller details of the person's life. The way she loved reading her Bible. The way he never drew attention to himself. The father who was always in the front row at his daughters' dance recitals. How her favorite thing was to go to Yankees games with friends. How, before he passed, he and his wife had an after-dinner tradition of strolling the neighborhood together hand in hand and talking about their days. The profiles gave a vibrant glimpse of how each person showed up for their life while they were here, while they had their precious sliver of time on earth.

I still think about those profiles. I think about how life is short and fleeting, how it's so easy to get caught up in things that don't matter. And then life snaps you awake. I think about how I want to look back and say, "I was here. I was present and I was here."

I want to see evidence that I lived this life well. That I didn't waste time. That I loved extravagantly. That I made people feel seen and added value to their lives. I want to have explored new cities and traveled to new places—if time permits me to do those things. I want to know that I ate nourishing meals and gathered people around my table whenever I could. I want to know, unmistakably, that I tasted and saw that the Lord was good—in the high and low moments. I want to look back and know, without any doubt, that I rode this thing until the wheels fell off.

But why does that feel like such a tall order?

I think it's because we often find ourselves straddling the line between living in the moment—really living in it—and documenting the moment. In our world today, the two sometimes feel inseparable. We all have our phones out. We all want a selfie. We all want proof that we were here, even at the cost of not being fully in the moment at hand.

The other day, I took our two-year-old daughter, Novalee, to the aquarium, and we were waiting for a cheesy little sea lion show to start. In front of us, four boys huddled together around a phone. They weren't talking to each other. They were ogling the device in front of them. When the sea lion show began, the boy at the end started recording the presentation. Novalee watched sea lions barking and sliding across the platform as trainers fed them fish, and I watched this young boy crane his neck and record the whole thing on his iPhone. He never once looked up at what was happening right in front of him. I couldn't stop thinking about that boy for the rest of the day. I wonder if he ever rewatched that video or if the sea lions still live in perpetuity in his camera roll.

I recently led a roundtable discussion on presence, and I opened by posing a question to the group: "Where in your life do you struggle to be present?"

One of the younger guys in the group raised his hand and said, "Can you give us an example?" A bunch of other people nodded their heads in agreement. "Like, what's an example of presence?" he continued.

Lane and I talked about that moment on the drive home, how it never occurred to us that people would no longer know what it means to be fully present. That they would need a definition and examples.

"I think I get it," said another younger guy before I could give a definition of presence. "I struggle to be present when I'm out somewhere doing something with my friends or on vacation.

I feel torn between staying in the moment and documenting the moment. There's this thing within me that says I have to document it and tell people I did it, or else does it even matter that I was here?"

The people in the circle nodded their heads.

He paused briefly and then said, "I find myself stepping out of the moment every time I try to document it. So, yeah, I'm torn."

Unplugging is helping me face that exact tension—what was really happening within me as I spent more and more of my time documenting and sharing my everyday life.

I was living less.

I was performing more.

I was struggling to find the balance.

Lane and I would go on a date night, and I would document the plates, the restaurant setting, or the big bag of popcorn between us. While capturing those things wasn't wrong, I realize now that I wasn't in the moment. I wasn't even digesting the moments myself before sharing them with others. I started to notice this even more with my daughter. She'd do something silly or sweet, and I'd whip out my phone to record her. The spontaneity of the moment would be crushed by me in her face coaxing, "Go! Do it again!"

It's harmless, you may say. But the documentation started to seep into every area of my life. And then it became performative. I would document the quiet time but not be fully present in it. I would record the date night and then argue behind the scenes.

At some point, I no longer knew who I was documenting things for—but it wasn't for me. An insecure voice inside me was seeking attention. My documentation was less about sharing and more about my need to build a false sense of self-esteem. I've always been scared to be overlooked, so with each piece of content I posted, I could say, "Look at me. I'm doing this thing. I'm living this life. Don't you see it? I was here."

When I hit my hundredth unplugged hour, I bought a little point-and-shoot camera to celebrate the milestone. I tend to be overzealous when I uncover a new passion—I immediately want to buy every accessory associated with said passion. This time I waited to see if I would embrace the unplugged journey. When the hours started stacking, I bought the camera.

It sounds dramatic to admit this, but the moment I held that camera, I remembered why I first fell in love with the art of documentation.

I was the kid carrying the massive ten-pound video camera on my shoulder back when making a home movie was a big deal. I was the girl sitting in front of the digital camcorder in high school and reading angsty poems and diary excerpts to the camera as if it were a confessional booth. I was the teenager at parties with the digital camera, capturing fragments of the night and then staying up until two o'clock in the morning uploading 5,603 photos to Webshots in those pre–social media days.

I have always loved documentation, and I always will. It's one of the reasons I still love and stay on social media. I think it's beautiful that we can enter into other people's days—peek behind the curtain and see what their morning routines consist of, how they balance motherhood and work, how they cook a Michelin star meal, or what their creative processes look like.

Parts of me blossomed as I used that tiny camera. I started taking candid shots at gatherings and gifted them to my friends—beautiful, raw images of their kids playing. I documented a gender reveal party without the couple even knowing it. I carried that camera around and interviewed everyone at the party. I learned how to edit videos and created a little keepsake for them.

I started printing photos and making photo albums. I love watching our daughter flip through the album pages on our coffee table. She has no idea she's growing up in a world where albums like these are quickly becoming ancient artifacts. Though I documented those moments, they still sit vividly in my memory, and never once have I felt pulled away from the present moment when using that little camera. I don't know exactly why that is, but I think it's because I can't instantly share the photos. The memories get a chance to sit and simmer.

And perhaps the most significant shift: I finally allowed myself to be in the photos. I stopped worrying how I would look in all the images and started handing the camera off to my husband. It became less about how I looked and more about how I want my children to have pieces of me—tangible evidence of me to hold on to—when I'm no longer here.

One would never think a small camera could expose and heal so much.

———

I'm sharing less these days, but I feel more whole.

Nowadays, I find myself pausing before I go to share something. It's almost like a vetting process. I ask myself:

What's my motive?
Am I sharing because I want others to join in this moment
    with me?
Am I sharing because I need to feel seen and validated?
Am I sharing in the hopes that someone will laugh? Or feel
    seen? Or find joy?
What's the motive?

I pause long enough to find one. If the motive comes back pure, I move forward. If it's rooted in insecurity, I take a step back. As a side effect of taking that pause, I'm learning to appreciate what's right in front of me, learning to show up fully and embody the moment at hand without trying to capture and keep it—without needing to share it to find value in it.

My little camera holds maybe a hundred home videos that we haven't watched yet. But one day, we will. And something feels sacred about the fact that they're just for us. They're for keeps.

But many more instances in our daily life don't get documented at all. These moments happen more and more these days. The phone stays put. The camera stays in its case. The full-body presence still feels jarring sometimes, like I need to grab something to feel steady. But as I practice presence, I remind myself to sink deeper into what's happening all around me—to stay in each moment a little longer.

When you experience a beautiful moment, you don't need to capture it or bottle it up. Feel it. Enjoy it. Drink it in. It's enough just to be there. Enough, enough. Your presence is your purpose in this moment.

chapter nine

SCREEN TIME

For as long as I've been a mom, I've seen the same constant debate playing out all over the internet and across momming circles: screen time. How much? How little?

At first, I was hyper-focused on it, but the more it came up, the more I felt like I had to shift the question back to myself: *What's my screen time like? Too much? She's watching me. She's getting her cues from me. She's figuring out how to navigate the sweet and salty of life from me.*

The average American picks up their phone 352 times a day. That statistic is up four times since 2019.[1] That's approximately every three minutes. What messages am I sending my daughter when I pick up my phone 352 times a day? What messages am I communicating when the shiny brick of technology on the counter gets more of my attention than she does?

Our daughter was just over a year old when I started the unplugged hours. We'd gotten into the habit of scooping her up from her crib in the early morning hours and bringing her into our bed for a few slow moments before the day picked up. We'd sip our coffee, talk a little bit, and inevitably reach for our phones to either check something or document something she was doing—a giggle, a babble, or an unexpected milestone. One moment we'd be completely present; the next, we'd be swept up in a social media rabbit trail. I would find myself holding my device and not even remembering why I picked it up in the first place.

During those slow mornings, Novalee started reaching for the phone, wanting to hold it herself. And why shouldn't she? She wants to mimic everything we're doing. She wants to hold what we're holding.

I started to ask myself: *What is the story I'm telling her about this device in my hand? And is it a good story?*

To be honest, I didn't want to write about motherhood.

Bringing up motherhood would feel more natural to me if we could be sitting across from each other having coffee, maybe at a small table in the corner of some cozy bistro. We could talk honestly about the struggles and triumphs of motherhood. We could be real about how it feels like we're constantly walking a tightrope where one moment we're swaying to the side of joy and elation and the next we're racked with anxiety over wanting to make sure we're doing this motherhood thing right.

When we bring the motherhood conversation online, we often allow many voices into the conversation in our quest to do things right. That small table where we once could talk so candidly becomes swarmed by dozens of others pulling up digital chairs. A second cousin enters the chat. That girl from high school we were never friends with but followed online makes a comment. We read another book. We listen to more podcasts. We seek advice. We search the trenches of motherhood forums. The table is suddenly cramped.

And while many of these voices can be valuable and make us feel seen, there are also plenty of volatile ones. Motherhood on the internet these days is a gauntlet. I never would have guessed this before having a baby, but the fangs come out on social media whenever the topic of parenting comes up. Let's just say, it's not always the friendliest of voices hanging out in the comments section. And if we're not careful about sorting through the noise, we can start to carry the critical voices with us. We load more and more of them into an already heavy bag of parenting to-dos slung over our shoulders, and we try walking forward while all the voices are screaming and debating:

Does your child know a second language yet?
Could they read coming out of the womb?
How much screen time are you allowing?
Has your child hit that milestone? What about that one?
Did you breastfeed? What about red dye 40?
Are you doing enough tummy time?
*Are you homeschooling? Are you working? Are you doing
   enough?*

Motherhood is hard enough without feeling like every social media post is a gauge for what we're doing right or wrong on the journey.

———

The more I unplugged, the more I learned to mute the voices—the constant opinions and the criticism. It wasn't an intricate process; it was simply the realization that less time online equaled fewer opinions flying in from every direction. And in the process of paring down the voices I let in, I discovered something surprising: The loudest, most critical, most shame-inducing voice I was hearing didn't belong to someone on the internet. It wasn't the voice of someone in my inner circles. It was me—I was the one putting so much pressure on myself.

I walked into motherhood like it was something I needed to accomplish. Before my daughter was even born, I plotted our future successes. My child would be literate, cultured, bilingual, and a brilliant artist—all by the age of two.

When she could take in the world around her, I started planning extravagant field trips.

When she said her first word, I stocked up on workbooks at Target.

When she figured out how to crawl, I started making sensory bins I saw on Pinterest, filled with pipe cleaners and cotton balls.

When she held a crayon for the first time, I was ready to show her how to stay in the lines.

Before you say it, I'll say it for you: I was both merciless and exhausting as I ran my little baby regime.

I was convinced I had to sculpt little shapes out of food and throw her a first birthday party that rivaled my wedding. But when I stepped back to sort through the voices in my head, I asked different questions: *Am I doing these things for her? Am I doing them for me? Or am I doing them because I want to prove myself to people who consume my life through clips, snippets, and photographs?* We can show up for all the wrong audiences without even realizing it.

Let's just be honest: The field trips typically ended in meltdowns. The workbooks are stuffed in a closet somewhere. The sensory bins ended up all over the floor, and I spent days picking up cotton balls and tiny beads. She still won't eat cucumbers, even if they're shaped like stars. Just yesterday, she gagged up a carrot when I tried to turn dinnertime into a "Choose Your Own Adventure" theme. And she'll never remember that first birthday.

The truth hit me one afternoon while I was sitting undistracted in the carpool line to pick her up from preschool. A tiny voice of grace entered the angry dialogue I was having with myself about how I could be doing this whole motherhood thing better. It piped up and said so clearly, *She doesn't need all of this. She doesn't need extravagance. She doesn't need you to perform or try to impress her. She doesn't need the elaborate field trips. She just needs you. That's what she wants, Mama. Your presence.*

The voice felt familiar, like I'd heard it before.

I know it was there when she first arrived in this world—those first few hours. I remember staring at my daughter after the nurses wheeled us into the recovery room, thinking: *Wow, you're here now. This is it. We're going to have to figure this thing out together, huh?*

As I gazed at her, I felt overwhelmed by what I can only describe as a sense of quiet capability stirring within me. It was just her and me locking eyes. The rest of the world couldn't get in. I looked at her with so much fear but also so much love. The love won out as I whispered, "It's you and me, baby. I don't know all about being a mother, but I've got you. You're my girl, and we're doing this thing together."

I think we could call that quiet voice intuition—and I think we all have it in us. It's a gift from God that can be honed and developed. But many other voices can get in and drown it out. We stop trusting that inner sureness that once told us so clearly, "We're okay. We're doing okay."

While many areas of my life are gradually improving as time goes on and the unplugged hours stack up, motherhood is an area in which the results were instant. Immediate. I can say confidently: The unplugged hours have made me a better mom.

Novalee and I started to uncover what I like to call "pockets of presence" throughout our days. The phone is away and we're connecting, just connecting. We build castles and parking garages. We sip imaginary tea from wooden teacups. We trace letters and we make art. I notice this brilliant and intense look in her eyes, like she can't believe I'm down on her level and giving her my full attention. We bake cupcakes or just make disasters out of the cupcake liners. We swim and we chat, and we walk around the block waving to all the neighbors. We schmear a glittery substance that resembles makeup but is definitely not makeup all over our eyes and cheeks. We paint nails.

It's mundane, sure, but I know it matters to her. I know it signals to her that I want to be down on the ground beside her— that there is nowhere else in the world I would rather be than right here with her.

I know she will grow up in a world saturated with technology. I understand that I'm her first teacher in this arena and that I can model behavior for her. To show her the importance of presence, I put away my phone and meet her in the moment at hand. Sometimes it's short. Sometimes it's longer. But there hasn't been a single pocket of presence from which I didn't walk away feeling lighter, surer of myself as a mom.

The other day, we sat on the floor and did a puzzle together. We traced her letters at the kitchen table. After Lane got her ready for bed, she returned downstairs in her bright yellow nightgown, her wet hair piled on top of her head, and she looked me in the eyes before saying, "Thank you for playing with me today, Mama."

I cried when those words tumbled out of her mouth—they felt so tender and kind. And they came from just a few minutes of sitting beside her giving her my undivided attention. A few kisses, a few high fives, a few connection points. It's not always that simple, but that day it was. That day, the presence filled us both to the brim.

chapter ten

WILD AND PRECIOUS

I knew I wanted to marry Lane on our third date.

Between our second and third dates, I'd tried to back away. He was good and kind, respectful and bubbly; he opened car doors for me and wasn't playing any games. In all my previous relationships, I'd been looking for fireworks, but this man was a slow-burning fire. He was a gentleman in all forms and fashions of the word. I look back and realize it was fear that made me want to walk away. It's easy to talk yourself out of good things when you're not sure you deserve them.

But my friends came around me and told me the truth: *This is a good thing. He is a good man. You deserve goodness. Don't run.* So I said yes to our third date. He asked to cook me dinner at his condo. "You can bring dessert," he said.

I walked into his industrial loft with the high rafter ceilings, holding a box of brownie mix. On the cooking front, this pretty much defines our relationship.

That night, he prepared scallops on a bed of orzo pasta. And while I know this sounds impressive, I learned that evening that he loves the challenge of a new recipe, whereas I can't even be bothered to follow one. He loves to chop, dice, simmer, and sear. The more steps and ingredients the better. I love the following steps in regard to food: add to cart, tip the delivery person.

We sat out on his patio, big plates between us. I was a recent transplant from Connecticut, and the chilly bite of the October air that night reminded me of home, of places where October is its own season. I don't remember much of what we talked about that night. But I remember how it felt—like peace and joy and calm were all mixing together. Like this was right, and this was for me, and it now made sense why other things weren't for me in the past.

After dinner, we prepared the brownie mix together and watched *Garden State*. There was a moment, as we waited for the brownies to bake, when he and I stood on opposite ends of the kitchen island talking casually, and I looked around. I'd borrowed a pair of his woolly socks. Dishes were in the sink. The recipe he'd printed at work lay on the counter. A bottle of red sat on the island between us. And for the first time in any dating scenario I'd ever been in, I thought, *This could be everyday life.*

I'd done the impressive dates. I'd done the fancy restaurants and activities. But what I was looking for was someone to partner with. Someone to live the ordinary Friday night with. A relationship where, when we stripped away the plans and the events and the parties and the extravagant, we were left with just Friday night together.

Seven years later, it's an ordinary Friday night. And an ordinary Friday night in our home means one thing: Pizza Friday.

It's a tradition we began as a family a few months into the unplugged hours. To know me is to know I go hard when it comes to traditions—declaring with a vengeance that we will now do this new thing for the rest of our lives—and it's no surprise that some don't stick for this very reason.

But Pizza Friday stuck instantly. I think that's because the recipe is so simple: add a day of the week, a pizza (handmade or ordered), and a movie. Combine and enjoy. We power down. We turn off the lights. We disregard bedtime. We snuggle on the couch, and we watch a movie all together.

Lane has fallen in love with the art of making pizza. We've discovered new pizza joints in our area. We alternate weekly on the movie choice. We've all bought into Pizza Friday. The sight of dough slowly rising and the aroma of cheese bubbling over in the oven are all the indicators we need to know the week is over. It's time to get off the grid.

Pizza Friday often gives way to a powered-down weekend for me. I turn off the phone. We sleep in. We spend time with friends. We cook a meal together and prep for the week ahead. We head to church. Ordinary life was happening before we started doing this, but I was too distracted to see or savor it. Before the unplugged hours began, I was always doing a few things at once. I was splitting my presence with emails, notifications, and idle scrolling.

We do these things because we're convinced we'll miss out if we don't juggle. It's classic FOMO. Author Oliver Burkeman writes,

> "Missing out" is what makes our choices meaningful in the first place. Every decision to use a portion of time on anything represents the sacrifice of all the other ways in which you could have spent that time, but didn't—and to willingly make that sacrifice is to take a stand, without reservation, on what matters most to you.[1]

Our no to one thing is often a yes to something else. Powering down can be a physical way of saying yes to the life right in front of us. Because the only alternative is to keep up the juggling act. It's impressive at first. But alas, some balls are plastic, and some balls are glass. It might be time to decide which balls we can't afford to drop before something truly precious shatters.

————

I asked Lane the other night what he likes most about the unplugged hours. Without hesitation, he answered, "The traditions that have come from it."

My focus is often consumed by internal shifts and ways of bettering myself, but unplugging has many external rewards. We're sitting at the dinner table together more instead of eating our meals in front of a screen. We're reading books at night after the plates have been cleared away. We're talking more. We'd been talking before too, but now we enjoy a different kind of undistracted, back-and-forth banter. Like we're getting to know one another again. I'm convinced conversations are different when phones aren't on the table.

We've gotten into the rhythm of practicing *hygge* together in the evenings. An article in the *New Yorker* defines the Danish term as "a quality of coziness and comfortable conviviality that engenders a feeling of contentment or well-being."[2] The word is said to have no direct translation into English, though "cozy" comes close. The Danes, famously known as the happiest people in the world, view *hygge* as their way of life—putting togetherness at the forefront of all things.

At first, we had no idea that what we were doing could be classified as *hygge*. It was my mom who pointed out the Danish quality to our nightly ritual. We clear the dinner plates, tidy up the space, light candles, and dim the lights. I often put on classical music, and we pull out the blankets and the books. It's a way to transition away from bright screens, notifications, and the pings of email coming in. The ritual moves us into a space of togetherness and coziness.

Justin Whitmel Earley, author of *Habits of the Household*, writes, "One of the most significant things about any household is what is considered to be normal. Moments aggregate, and they become memories and tradition. Our routines become who we are, become the story and culture of our families."[3]

We've taken this idea to heart in our family, but here's the thing: I won't airbrush the picture so it looks like we are perfectly in sync and every night is intentional. Often, it is a fight against all the forces within us to get to the dinner table or to remember to pause and see one another in the chaos of our busy days. We don't always light the candles. We don't always get along. But we've decided that these things matter to us, so we pursue them. Some days we only sit at the dinner table for ten minutes, but it's a stake in the ground; it's our family showing up to say, "This matters. We're fighting for it."

———

In my twenties, I didn't think much about traditions, family values, or things that take time to build. I wanted life to unfold like a movie. I chased the good stories at all costs. The person I was a decade ago believed life needed to unfold with all the drama of a Taylor Swift album. Former me felt a constant pressure to seize the day and make each moment mean something. For her, meaning was the biggest thing. Everything had to *mean* something.

I'm afraid that girl was wrong. I'm just now getting to a place where I can finally say that not every hour and every moment has to mean something. Sometimes moments are just moments. We don't have to assign worth or meaning to them. They're simple and beautiful, and they leave as quickly as they arrive. We can't catch them or bottle them up and hold on to them forever. We're not supposed to. We don't even have to try.

But that girl in her twenties would be shocked to know that I've encountered more magic over the last few years than ever before. And that magic exists mainly in the ordinary. There is so much magic in the everyday walking-around rhythms and routines of daily life.

I'm finding out that our everyday lives are something we build. We get to decide where the magic lives. And when we start to look for it, we find it everywhere.

Surely you will find it too. It will appear in the sound of a baby's heart beating against your chest. It will emerge in the scrubbing of dishes as you realize this is the first time in such a long time that you've felt the water and suds on your skin. The magic will arrive as you walk outside to get a glimpse of the horizon instead of checking your phone, or as you sit out on the porch while the cicadas hum, waiting to examine the flowers that bloom only once the moon is up.

Building a life requires being present to what's in front of you and having the courage to start changing things. It requires the willingness to tune in when you'd rather tune out.

Building a life requires letting go of what didn't happen—the paths you didn't choose. The people who weren't for you. It means being brave enough to stop looking in the rearview mirror—nothing is waiting for you there.

Building a life requires laying down the ideals other people have for you—the way they want your life to be. It requires taking the time to carve out values and decide what you're fighting for.

Every day, your choices reveal what you're fighting for. Whether that's presence or togetherness, wellness or the things of God—what you do with each day's allotment of hours will be the makings of what poet Mary Oliver once described as your "one wild and precious life."[4]

You are allowed to sink your teeth into the life you're building. You are allowed to love it immensely without the fear of losing it. You can find deep pleasure in the ordinary. You can open your eyes to what is already in front of you. Yes, there will be mountaintops. Celebrate them well with people you love. But much of life is built from ordinary actions—and when you tune in long enough to be mystified by the ordinary, that's when it all becomes wild and precious.

———

On Valentine's Day, Lane gives me the best gift I've ever received from him. He has no idea that this is the case as he watches me unwrap the paper.

We always joke about how much his gift-giving has evolved over the years. Something shifted after that one time he gifted me a wine tote and a bright orange sweater that, since I was seven months pregnant at the time, made me look like a massive pumpkin. He explained that I could use these gifts when I wasn't pregnant anymore, and I cried. He changed strategies after that.

Years later, his gift-giving has come a long way, and I think it's because he's paying more attention. He's noticing the woman before him—what she loves, what makes her come alive. And, for the most part, those things stay the same: paper goods, books, coffee cups, more paper goods, more books, and coffee for the aforementioned coffee cups.

I peel back the wrapping paper to behold a set of cloth-covered journals with golden numbers on the spines. There are five of them in the collection. It's a five-year journey with a small space for each day to write down thoughts, events, and details of the day. I know this might sound like the most daunting gift ever; who would want to remember that much? But for someone who has been on a journey to savor daily life more and more? This is the next step.

I keep the journals by my bedside table. I've noticed a trend emerging in the pages where I've journaled the details of each day. The things I scribble down are never the big things or the monuments of success I've spent so much of my life striving for. It's the ordinary stuff. It's the stuff I used to miss when I was distracted by screens.

Mountain air. Girl talk. Another chance to unwind and unplug. Corn on the cob. Sitting out on the back deck.

Chicken fajitas. Gluten-free pancakes. Murder mystery games. Date night. Marvel movies. Long baths.

On an ordinary day in March, I write into the five-year journal, "A spirit of enoughness." That says it all.

chapter eleven

WE TRY AGAIN

We are currently in the phase of helping our daughter navigate all the big and complicated feelings starting to take shape inside her. I wonder if this process should really be called a "phase"—it feels everlasting, and I'm beginning to think we might be navigating this territory forever.

I often have to remind myself that each emotion she experiences is new to her. Joy. Sadness. Anger. Awe. She has no language to wrap around the feelings yet. When she feels an ache, she doesn't know to call it sadness. When she stands nervously on the edge of something new and uncertain, she doesn't know whether to name the sensation fear or excitement.

I once heard someone say that toddlers experience the same-sized emotions we do as adults, and yet their bodies are so much smaller. Imagine containing all those big emotions in such a tiny frame.

In case you're envisioning that our every interaction is a serene mother-daughter conversation full of gentle explanations about human emotions, let me pop the bubble for you. Just the other day, Novalee's sadness quickly morphed into her spreading out on the kitchen floor, limbs splayed in every direction—a full-on tantrum over the color of the water bottle she needed to drink out of.

The reality is that we're both undone by the end of most days. I know I'm supposed to be the compass holder, but I rarely know what direction it's pointing.

Lately, we've adopted a household mantra that applies to all of us: we honor all feelings, but we don't honor all behaviors. The feelings are important. They tell us things. They're indicators, nudging us to pay attention. But where the feelings want us to go—that's usually where things get dicey.

It's okay to feel hangry, but it's preferable not to snap at everyone within six feet of you.

It's okay to feel hopeless, but you don't have to let hopelessness convince you to lie in bed all day scrolling.

It's okay to feel all your feelings, but throwing tantrums on the kitchen floor won't get you anywhere.

In our house, when we are spread-eagle on the kitchen floor, we defer to what most of us need when our emotions run high: a time-out. A chance to breathe and get ourselves back together. And when I say time-out, I mean we all take one. We all, collectively, take a step back to breathe and return to center.

After a few minutes, I walk into my daughter's room to have a conversation. At the end of the conversation, I ask her the same question I ask every time: Are we ready to try again?

She nods her head and says, "Yes, Mama. Try again. We try again."

She and I both feel a moment of redemption when that question is posed: Are we ready to try again?

She agrees to it. I agree to it. We clear the slate.

———

I'm not a parenting expert, nor do I want to be. Please don't send me emails with advice or tell me how time-outs will make my child grow up to rob banks.

But here's what I do know: The more I parent, the more I learn just how much "parenting myself" is part of the job. My daughter and I are both learning there is room to try again. It's quite the revolutionary concept for a recovering perfectionist. Each time I tell myself that we can try again—that we get another shot—it feels like a cool drink of water to the parts of me that have been thirsty for grace for so long.

When I get down at her level to tell her this, I'm telling it to myself too: "Our next steps don't have to be perfect. There's nothing wrong with messing up. We can always try again. The grace is plenty in this place."

These words floated back to me one evening as I was sitting with my friend Hayley at her dining room table. I was a couple months into unplugging, and we were talking about how the journey was going.

Hayley and I couldn't be more different, and that's one of the many reasons I love being around her. I often pose challenges to myself—new rules and ways to improve myself—and Hayley just gives me the same exhausted but kind look, as if to say, "I love you so much, but you also don't need to do another challenge."

So I was barely surprised when she said nonchalantly, "Yeah, but what about the days when unplugging doesn't happen? There are days when I end up on my phone way more than I want to be. What then?"

Same. I have those days. When you start unplugging, your life doesn't morph into a perfect oasis where your phone is always magically off at the right time. Some days are really easy, and some days, despite our best efforts, we fall into a black hole of scrolling and online shopping late into the night.

At first, this frustrated me. I would have blissful days when the unplugging came so naturally and the hours stacked up—and then I would go two or three days feeling like the phone was glued to my hand. I couldn't understand why this practice wasn't becoming an ingrained habit more seamlessly.

That's when I had to apply some grace to the process and realize I'd been cultivating my plugged-in habits for years. There had been years of scrolling on social media before that first cup of coffee. Years of checking email in checkout lines. Years of falling asleep to the light of my phone plugged in beside me. Years of checking out when I could have checked in with myself. Years of clicking news articles between tasks at work.

All these habits, left unchecked, became a subtle addiction.

Little by little, we've all become more plugged in. A habit begot a habit begot a habit. If we want to undo some of these habits, it will likely take some time. It will take some fits and starts. It will take days when we feel like we should rip up the unplugged hours tracker and throw it in the garbage. But instead, we can recommit ourselves to the reason we likely started powering down in the first place: the chance to be present for the life unfolding right in front of us.

"Those days happen," I assured Hayley at the table. "Absolutely. But that's when we wake up the next morning and ask ourselves, 'Are we ready to try again?'"

As long as we're willing to try again, there's hope.

I've tried to craft my words here carefully, because I don't believe we need any additional pressure—we have enough already. Most of us are already wringing our hands with guilt over how we could have done that one thing differently six years ago.

And mothers, especially, seem to get an extra dose of that pressure. There's a whole slew of online messaging about how we only have approximately fourteen summers left with our children, 5,840 days left until they move out of the house—equating to only 140,160 hours. *Use them wisely! Be present! Be grateful! Don't you dare miss the chance to get down on the floor and play Magna-Tiles! You will miss these days!*

I know this kind of content is meant to jolt us awake, to make sure we savor the time at hand, but I think it can unintentionally stir deeper feelings of guilt and shame within those of us who already feel like we're doing the best we can, and it isn't always enough.

Yes, time is going by quickly. Yes, we want to press in. But there's grace—all the grace—for the days when we miss the opportunities.

And grace for the days when we feel like we've failed ourselves or others.

And grace for the days when we're trying to be present, but we're just not connecting. Our phones might even be away, but our mental to-do lists are stacking up; the disapproving voices are hitting us from every angle.

This journey isn't about perfection; it's about learning as much as possible along the way. Even those days when we feel mentally checked out can be subtle teachers.

———

Last month, my mom stayed with us for a few days, and she and Novalee went puddle-jumping one day after an afternoon storm cleared.

The two stood ready at the door—Novalee had donned pink unicorn rainboots. Her mini polka-dotted baby stroller stood beside her, the ragged plush Elmo the third strapped inside.

I tagged along on the excursion, not because I wanted to but because I felt a sense of guilt that told me I should. There's that pressure again—the relentless nagging in my brain to not miss out on the "moments that matter."

What followed was one of the slowest, most grueling activities of my entire life. Like, unreasonably slow. We stopped and started every few seconds because Novalee insisted on pushing the play stroller at an excruciatingly slow pace.

Whenever we came across a puddle, Novalee and my mom would stand over it for ten minutes, surveying it as if it might prophesy to them. I marveled at how they seemed to have all the time in the world to stare at this murky pool of water. When the actual jumping in the puddles began, it went on forever. You could not have convinced these two they weren't at Disney World at that moment.

Novalee and my mother are kindred spirits, and I hope that never changes. I hope my hurried pace and my general discontent with sitting still never rub off on my daughter, that she can hold on to the beautiful slow spirit within her—the one that knows to stop and savor and delight in dirty rainwater.

The moral of this story is that I should have stayed home and left the puddle crusaders to their excursion. I was pacing back and forth the whole time, exasperated by their leisurely approach to the outing. My mood was sour and palpable, and the message I was sending was clear: *I think I could be doing something better with my time right now.*

I continued to think about the puddle jumping long after my mom boarded a plane and returned home. It wasn't that I felt guilty or wished I'd enjoyed the moment more. It was more that I felt like I'd missed something. I wanted a chance to try again.

Today, it's raining again. It's mid-afternoon and I'm sitting at the computer, checking the forecast. The sun will be out again in another hour. Because of the heat, the puddles will dry up fast.

I look outside the window at all the puddles. Novalee has been talking about them all morning. I already told her, not today—my list feels too full for me to spend time searching for puddles after the storm. But sitting here, caught between a list of tasks and what my gut is telling me, I know what I need to do.

I enter her room, nudge her awake from her nap, slide her pink plastic jellies onto her feet, and bring her outside. I leave the phone behind on the counter.

Together, we set out into the rain.

Instantly, she comes alive. Her mood is electric.

It's just us. The rain spitting in our hair. And puddles galore.

We explore the neighborhood. We visit the geese at the pond down the hill. I watch her run wildly down the paths. I wish I could snap a picture of her now—looking so free and untethered. Instead, I take a mental picture. For me. The memory will be enough for me.

As we climb back up the hill, soaking wet and hand in hand, I think about how this is one of the more magical moments I've had in motherhood. And it all came from a do-over. From being able to say, "I think I missed something back there. I'd like to try again."

And so we try again.

We can always try again. The grace is plenty in this place.

chapter twelve

WALK IT OUT

———————

My earliest memories of my mother are of her sitting in the same spot at the kitchen table every morning. A cup of coffee would be beside her, and her leather Bible laid open. I can still picture her sitting there—the sun pouring through the kitchen window and flooding into the small dining area where she'd be scribbling into a plain one-subject notebook she bought for ninety-nine cents at Walgreens.

Years later, I would find out she was copying the Bible during all those mornings. Word for word, letting the text seep deeply into her mind as she wrote it all down. She'd fill a notebook and then throw it out. The part of me that has never thrown out a notebook cannot fathom this. But she would say the notebook served its purpose, and it was time to get another. If she'd bothered to keep the notebooks, she'd have well into the hundreds by now.

When I was a child, I watched her morning rhythm curiously. When I was a teenager, I eyed her rhythm skeptically. I didn't understand God, but I was pretty positive I wasn't interested in getting closer. I'd always felt like an outsider when it came to matters of faith. But I also grappled with a sense of longing within me—a great hoping—that my life wasn't an accident. My mom's mornings at the table always made me wonder, *What kind of God is she meeting with? How good can he actually be? And what keeps her coming back there again and again?*

Our family walked through plenty of seasons in which I would have expected my mom to put the pen down and stop showing up at the table altogether. She didn't, though. If anything, she doubled down.

I never asked her about her practice, but I knew I wanted what she had: something or someone good enough to revisit thousands of times, no matter the circumstances.

And so, when I was exhausted from trying to fill the crater-sized holes within me, the parts of me longing for more meaning and depth, I remembered her. The woman at the table. The woman who never forced her faith onto me, handing me something I wasn't ready to hold. She never told me what I had to do or who I had to trust. She just walked out her faith—slowly, quietly, faithfully, consistently. Her daily life was, and is, evidence of decades spent at the table.

Today, my faith happens at the table too, though it looks different than it used to. Sometimes the voice of Elmo introducing the letter of the day can be heard in the background. Sometimes a live performance of *Encanto* is happening off to the side of the table. But regardless of the chaos, I show up at the table as much as possible because I want my daughter to see me filling myself up. I want her to see me walking out the practice of faith because, at one point, I watched someone walk it out too—and it changed everything for me.

———

A few months ago, I met up with a friend at his office, which resembles a nicely furnished fishbowl—with glass walls enclosing us on all sides. Anyone could look in and watch.

My friend works in the publishing industry, and we discussed the books I hoped to write in the future. He asked me what legacy I wanted to forge through my writing. What did I hope people would say about me over the next fifteen years? I was surprised at how quickly the answer came to me, as if I had been waiting for someone to ask me that question for a long time.

"I just hope people say I was the real deal," I answered. "More than anything, I want to be the real deal."

Tears unexpectedly filled my eyes. I can't tell you the number of things that had to shift around in my heart over the years to get me to this answer. I've spent years of my life being consumed by whether people liked me. Approval was my highest motivation for so long. Before starting the unplugged hours, I could see a direct correlation between how much time I spent online and how it affected other parts of my life. There was a disconnect between the life I wanted to be living and the life I was curating for others to see.

I'm convinced it's never been easier than it is today to project a false life onto a screen for others to follow. Especially when it comes to matters of faith, you can convince everyone that you're holding it together, that you're rooted deep in something other than yourself, and that you're doing just fine. If people don't look too closely, you'll pass without question.

Just last week, yet another pastor fell off the pedestal other humans had built for him. Building pedestals is nothing new. What's new is how we've gotten into the habit of waiting for people to fall from the pedestals we've put them on and then turning against them the second they fail to meet our expectations. I get it. I'm disappointed too. But I'm also not surprised it keeps happening. It's easy to get swept up into a performance.

The more we project our lives onto screens, the less we invest in what's happening off-screen—the threads of our lives that matter. How we treat others. How we pray. *If* we pray. How we show up. How we speak to ourselves when no one is listening. How we wrestle with the challenges of this world. How we take this complicated, gritty faith thing and walk it out day by day.

People always say we are what we do when no one is watching, but who are we if we give others access to watch us at all times?

———

So what does it look like to really, truly walk out your faith?

I don't have all the answers, but the more I power down, the more I find that walking it out is about investing far more energy into the practice than the preaching. It's like that well-loved quote often attributed to St. Francis of Assisi: "Preach the gospel at all times, and if necessary use words." Our faith—unspoken but lived out faithfully—may be the most powerful sermon we ever preach.

Walking it out means being willing to put ourselves rather than others under the microscope—to deeply examine how we think, pray, judge, and love. It means having the courage to change and accepting new challenges of transformation all the time.

Walking it out looks like getting things wrong—a lot. That's guaranteed. It means having people near enough to witness us getting it wrong and then love us back in the right direction rather than canceling us for being human.

Walking it out means continuing to show up and just keep walking. We take each day and do our best to move another step forward, and some days (read: a lot of days) we'll wobble. We'll fall short and still come back from it. No matter the number of falls, the goal stays the same: Just keep walking.

It's the end of the day. Novalee and I are working through her bedtime routine. We brush her teeth and change her diaper. I switch on her night-light. It's a small ceramic lamp with little stars cut out of the base. Light pours through the little cut-out stars, projecting a faint, glowy night sky onto the walls of her bedroom.

I sit in the rocking chair, and she climbs into my arms and lays her head on my chest. We rock.

For a long time, I had gone through the motions of bedtime with her as quickly as I could. I was often impatient, tired after a long day. It wasn't until I started leaving my phone downstairs that things shifted for us. We started taking more time together. We fought less. And my child, who never cuddled me before this new era, started asking to snuggle with me in the rocking chair.

Some nights we stay quiet together as I rock us back and forth. Other nights, we talk about everything she is planning for tomorrow. The things she will do. The dresses she will wear. I marvel at her—for me, tomorrow is a placeholder for my worries, but to her, tomorrow is only and always possibilities. Infinite possibilities.

Tonight, we sit in silence. I trace my fingers along the frilly edges of her princess nightgown. And I hear her little voice whisper into the darkness, "Wanna pray?"

I've talked to her about prayer in carpool lines and before meals. About how we can talk to God about our friends and family, and he's listening. In a world where we often fight to be seen, he stands waiting like a gentleman with infinite patience—ready to offer us his full attention. Tonight, talking to God is her idea. So much of motherhood feels like wandering around in the dark, hoping you are doing something right. Moments like this one feel like faint flickers of "Yes, you are" in the midst of the wandering.

She doesn't wait for me to answer her question. She starts rattling off a list of names as if she'd been holding them in her mind all day, waiting to release them: *Pop-Pop. Abuela. Skipper.*

"Who else? Who else?" she asks every couple of names. A phrase she's mimicking from me.

Nana. Daddy. Uncle John-John.

She keeps whispering names into the darkness. I draw her in closer to my chest and tell myself the truth as we rock back and forth: *Yes, we're walking in the right direction. We're stumbling, sure, but we're making it—one wobbly step at a time. Hold tight to the glimmers—they're shining in plain sight. Hold tight to the glimmers and just keep walking.*

chapter thirteen

BEGIN

I belong to the population of people who fully believe we are one Peloton bike away from changing our lives.

We are the ones easily lured in by ads. We are highly, highly susceptible to multi-level marketing. We buy all the creams and the powders and the serums. We often feel like the latest thing being advertised—an ice roller, retinol serum, or a new workout app—will be "the thing" that wipes away all our problems. It will be the product that finally propels us into a new and improved way of life.

Who knew a cast-iron pan could be the secret to happiness and abundance? Me. I knew that because I consumed the ads, imagined my family gathering around the pan as if it were the fountain of youth, and saw the life-changing pan in my browser enough times that I finally bought it. Life change status: pending.

That's what constant connectivity paired with over-consumption is doing to us—it's getting us to believe a narrative about how that thing in our Amazon cart will be the thing that finally brings progress into our lives. We'll be changed. We'll be new.

I see this all the time with the students I coach in writing. The most common reason they don't start is because they're convinced they need something outside of themselves before they can get started. A new writing software. A different notebook. Fresher ideas. I tell them the same thing I tell myself whenever I begin writing: You don't need another thing. You need to sit down and start with what you have. What you have is enough.

It's easy in this digital age to get swept up in other people's momentum. Even as I unplug more and more, I'm aware of how quickly I can be influenced by what seems to be working for someone else when I log onto social media or open my inbox. If the results I'm getting aren't instant enough, and someone else's journey looks more seamless, I'm tempted to exit the lane I'm running in to follow them. Do that enough times and you'll find you're not actually moving forward; you're moving side to side and wondering why nothing seems to be working for you—why the weight isn't coming off, why the book isn't being written, why the growth isn't happening. One powerful thing I've been learning to do in this season is to see someone else having amazing breakthroughs and say, "That's amazing that things are working so well for them. I'm going to stay the course, though. I don't want to miss what's happening in my own lane."

Psychology tells us that watching others can be a motivating factor. But at some point, we start to confuse information with action. We convince ourselves that a steady stream of content, blogs, podcasts, sermons, coupon codes, reels, products, and videos will give us everything we need to grow and transform. However, the truth is that all that information flying at us from every angle is immobilizing us. It's holding us back. We keep buying the products, the courses, the programs, the apps. But we don't bother to take that first shaky step. We become so busy spectating, watching from the sidelines, and convincing ourselves we need what others have that we never begin.

It's easy to romanticize what other people have, the things they've put time and sweat into earning. It's easy to want those things for ourselves. But meaningful progress has never been a spectator sport. Technology may advance rapidly, but AI can never step in and live our life for us. That part is on us. I don't want us to miss out on the growth and transformation that could have been ours because it was "good enough" to watch someone else write the book, go back to school, get healthy, or cultivate a rich and deep faith life.

And while part of me wishes I could show up at your door and blast into your room with a megaphone like one of those over-the-top coaches on reality television, I know that's impossible. Pep talks only go so far, especially when you're terrified to change or grow because you've gotten comfortable with not moving. I can't be the one to convince you that watching others is not enough. I can't get you to stop buying all those things that promise to change you. You have to get to a point where you come to the realization on your own, where you tell yourself, "It's not enough anymore. It's not enough to watch. It's not enough to make excuses. I'm going to step into my own progress."

I saw my friend Phil at a coffee shop the other day. It had been months since I'd seen him, and I couldn't help but notice a striking difference. He looked so healthy. He exuded a vibrancy I hadn't seen before, and I had to ask him what he was doing differently.

He proceeded to tell me about a dusty set of barbells he'd found in his parents' basement, along with a bunch of other old gym equipment that once belonged to his grandfather. He'd been wanting to move his body more and establish some better habits, but he kept telling himself he needed an expensive gym membership to make that goal a reality. And then one day he grew tired of telling himself that story. He transported the gym equipment to his house and, with a yoga mat set on the grass outside along with those dusty barbells, he began working out. He decided he could build from there. He went off the grid for a little while. He pressed into his progress. New habits took form. And he discovered a new way of living was waiting for him.

I love the image of the dusty barbells in the basement. It's another reminder: You and I don't need yet another "thing" before we can show up today. We don't need another gadget. We don't need that new app with the seven-day trial. We don't need that workout program or those microgreens the influencer is peddling on her story with a 20-percent-off code. We don't need another self-help book or podcast episode.

We don't even need to tell people we're beginning. In the moment, it's tempting to post big, life-shifting goals and announce to our aunts and high school friends on Facebook that we're about to change our lives, but there's compelling evidence for doing the opposite—for holding off on that big announcement and just deciding to start for ourselves. Telling people our goals online can lead to an instant flood of dopamine when the comments and praise start rolling in—tricking our brains into believing we've already accomplished the goal and diminishing our motivation to start.

What we need is already within us. It's sitting there, waiting to be acknowledged, like a dusty set of old barbells. We must stand up, remove distractions and excuses, and begin. Imperfectly. Shakily. Simply. Unsure. Afraid. We must turn down the volume of our culture and listen to the rhythm of our own aspirations telling us in quieter ways: Begin, begin, begin.

Begin

Begin writing.

Begin moving.

Begin picking up the camera again.

Begin stepping toward God.

Begin healing.

Begin mapping your way forward.

Begin powering down.

Begin moving off the sidelines and back into your life.

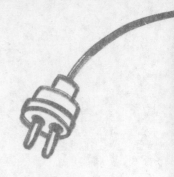

part two

SHIFTS

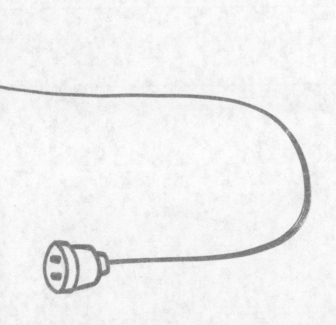

chapter fourteen

REFRAMING PRODUCTIVITY

I've always been a lover of productivity.

I love efficiency. I love tackling tasks. I love talking about tackling tasks. I love feeling victory as I plow through a to-do list. Each line struck through an item on the list fills me with a deep sense of satisfaction. Sometimes I even add tasks I've already completed just to cross them off and get that extra dopamine hit.

I realize not all people are like this. I'm not sure my mom has ever made a to-do list. She goes through life with whimsy and wonder, certain that whatever is meant to get done will get done in its time. My mom is also the most present person I've ever encountered, so maybe there's a tension between "presence" and the "need to get more done."

Most days, I'm the opposite of my mother.

I often felt overlooked growing up, which fueled my need to constantly do more to prove my worth. I'm the youngest of three siblings, and the only girl. One of my older brothers was a star athlete, and the other was that popular, cool, smart guy that every teen drama has. A lot of people were surprised to find out I even existed. I loved my brothers but felt invisible next to them—pinned in their shadows. I was already a bit of a recluse—always creating little worlds off in the corner—and I wasn't much to look at. Unruly red curls, a painfully shy demeanor, and a lanky little figure that earned me the playful nickname of "Boneyard" from my dad.

When I moved out of our small town for college, something clicked into place for me. I found the areas in which I could excel. And I think the younger version of me—little Boneyard—decided to battery-power the entire operation.

"Do more! Accomplish more! Say yes! One more thing—you can juggle it! We'll show them!" she roared as we added more things to our plate.

We'd get the 4.0. We'd tutor other students on campus. We'd run events. We'd intern. We'd do extra credit. We'd give campus tours. We'd be the prized pony. Boneyard was maniacal, and I loved having her driving me at all times.

———

Ambition can be a beautiful thing, but you can also start to fling your life away to a calling that never actually called your name. A calling to be all the things to all the people. A calling to look like you have it all perfectly together. A calling to have next to no margin on your calendar and always something pressing to do. The pressure is constant.

You begin tying your value to what you're doing—building an identity out of what you can manage to accomplish. You equate achieving more with being worth more—like you're a stock whose value goes up and down in dollars and cents based on what you manage to do today.

If you can do more, you'll be loved more. And for the starving, performative parts of you, that's all they've ever wanted: to know and experience more love.

Technology has only made things trickier for us achievers and people pleasers, because we can easily boil down our worth to likes and mentions and followers. If we share the ordinary parts of our day, not only do we receive affirmation, but that affirmation is also broadcasted to a wider audience. In his book about social media, *The Chaos Machine*, journalist Max Fisher refers to this broadcasting as "the most powerful form of approval because it conveys our value to the broader community." He writes, "When's the last time fifty, sixty, seventy people publicly applauded you off-line? Maybe once every few years—if ever? On social media, it's a normal morning."[1]

The parts of us that have always wanted approval—to be seen by others—take this digital applause and turn it into fuel to keep the engine running.

———

Productivity always meant one thing to me: getting stuff done. But that definition only works . . .

when chaos is at bay.

when babies are sleeping.

when distractions are minimal.

when life isn't doing anything unpredictable.

Shortly after I started unplugging, Novalee, who was eighteen months old at the time, became unexpectedly sick. It marked the beginning of a ten-month season of sickness.

It started with what we thought was a onetime bout of car

sickness coming home from a road trip. But that quickly morphed into car sickness every time we got into the car. We tried everything to make it go away. The sickness began showing up outside the car a few weeks in. She started getting sick first thing every morning.

Between doctor visits and hospital stays, we were forced to embrace a new routine that shifted by the hour. Some days were mostly normal. Other days were a wash from the moment we picked Novalee up from her crib. We'd take on the days in bite-sized pieces. Hour by hour. Moment by moment.

There were so many days when I was anything but productive by my own standards. When the tasks weren't checked off the list. When the emails went unanswered. When I didn't remember to drink even a sip from the obnoxious gallon of water sitting on my countertop. When the bed wasn't made. When the day was dismantled before the coffee had even brewed.

My definition of productivity—the one that had fueled me for so many years—was crumbling.

———

"I only feel valuable when I'm producing something and when I can prove to others I'm producing something." I write these words into my journal and stare at them during one of my morning quiet times.

I've operated for so long with a specific vocabulary.

Crushed it.

Killed it.

Got it done.

Hustled hard.

Did all the things.

I need a new vocabulary. These verbs aren't working for me anymore.

My mind flickers back to a memory from my first year of college.

I am eighteen, sitting on the bright purple bedspread of my first dorm room. My long-distance boyfriend is on the other side of the phone. We are breaking up, and it feels like slowly, painfully disassembling a house we'd lived in for years.

When there is nothing more to say, I call my mom. She drives nearly three hours to pick me up. We end up at a nearby Applebee's, and she orders a thick slice of chocolate cake—her remedy for all of life's heartbreaks. Actually, it's her remedy for nearly everything. Feeling lonely? Chocolate cake. Feeling celebratory? Chocolate cake.

I sit across from her—that big piece of cake between us—wringing my hands in my lap as I reason aloud: "I can fix this. I can definitely fix this. I can make him want me back. Or I can make a system of sorts. A system to make me better and stronger and wiser because of this."

"Or you could just be sad and not make a system," she says quietly. "You could maybe just eat a little cake and cry if you need to."

She nudges the white plate closer to me.

You could maybe just eat a little cake and cry if you need to.

All these years later, her words are still the permission slip I need. Permission to not hold it all together. Permission to not rule each day or "tackle" each task with mastery. Permission to put down the notebook and the planner, take off the running shoes, and, if I haven't taken a deep breath and looked around lately, just do that.

For so long, I've been hardwired to find my sense of purpose through doing, not being. I've strived for external benchmarks, shiny accolades, and robust task lists because they feel like proof. Proof that I'm okay. Evidence that no one has to worry about me.

Proof that I've "made it," that I've made something of my life. There's still a little Boneyard in me, after all.

But when I think about the person I want to be, that person is different from what I've been striving to be. I didn't realize it at first—not until the unplugging began—but I've been chasing after things that I don't even want anymore.

When I'm honest with myself, the person I want to become doesn't feel the need to prove herself all the time. She's patient. She's kind, not just to others but to herself. She listens. She's willing to drop whatever is going on to help someone else. She has an inner calm reminiscent of Psalm 1—about the tree that's planted firmly by the water and produces what is necessary for each season, without ever withering. More than anything, she doesn't need to accomplish to know she's okay. She's just okay, and her okayness goes with her into every step she takes.

I'm not that person yet, but it's possible I'm getting there. And I'm continually coming to grips with the fact that I can't produce my way into becoming her.

I'm telling myself a different story these days, even as items remain undone on all my lists. When I'm tempted to say nothing is happening, I'm choosing to believe something is happening beneath the surface. And as I cultivate that belief, my need to prove myself to others is beginning to chip off me like cheap nail polish. The process is so slow that it makes me want to throw Miracle-Gro in all directions. But I'm willing to hold out a little longer.

———

You will have "those days," friend. They're inevitable.

Those days when your plans crumble before you even walk out the door.

Those days when the babe is sick, and all you can do is cuddle them on the couch as they nibble on saltine crackers and watch *Frozen* for the six hundredth time.

Those days when the life you mapped out for yourself suddenly doesn't exist anymore.

There will be days of toddler vomit. Days of chaos. Days of crayons marking the white walls. Days of flat tires. Days of spilled coffee on white shirts. Days of traffic and days when you can't get out of your own way. Days when you stall, stall, and stall because even though you know the work you must do is important, beginning feels like taking a plunge into an ice-cold pool.

There will be days of these things and seasons of these things. And you might be tempted to say, "I'm not getting anything done." I've stood in that space before. I get it.

But what if we learned to look closer? What if we could get beneath the surface of the shiny exteriors we've built for ourselves and find something better? The days might not feel productive by our culture's standards, and maybe we'll have nothing external to show at the end of them, but I believe something unshakable and lasting is being produced in us as we keep showing up.

Patience. Preparedness. Trust. Gratitude. Assurance. The ability to be truly present in the moment. Could it be that on the days we feel like we're producing nothing, God is producing something valuable and resilient within us?

Through the pain. Through the doctor visits. Through the grief. Through the heartbreak. It could be that a better, slow-growing fruit is maturing for harvest right on time. It could be that something is about to bud or break through the surface that is far more lasting than any task we ever placed on a to-do list.

You might step out of this season as the most patient version of yourself to date. You might become more empathetic than you ever imagined. Maybe you'll finally find peace. Or maybe you'll find compassion for yourself bubbling up inside you.

You may become the kind of person who can comfort. Who can hold. Who can make the most killer cup of tea on a dark day and be there for the person sitting across from you. You might find your capacity expanding and your faith increasing as you learn to bend and surrender to what's out of your control.

You may finally become the person you always wanted to be before the world whispered in your ear and told you to hustle harder for your worth.

There is no productivity app for this kind of growth.

chapter fifteen

I AM ALWAYS LEARNING

W hen I first started writing on the internet, back in the pre-historic blog days—which I would boldly claim were the good ol' days because people were still relatively filtered, civilized, and kind—a conversation was emerging around the idea that you could carve and sculpt an online presence. You, the same human who burns eggs on the stove and forgets to use your blinker, could be a brand. A personal brand.

I met with a branding coach who helped me figure out who I wanted to "be" on the internet. She asked me what I was passionate about and how I wanted to make a name for myself. Up until that point, I hadn't given it much thought. I was twenty-three years old and had just moved from New York City back into my childhood bedroom in North Haven, Connecticut. Who could I honestly be in this big wide world?

The truth was, I just wanted to write good words that made people feel things. I'd found the book *Eat, Pray, Love* on my mom's bookshelf after I finished my first year of college. I was dealing with the aftermath of my first real heartbreak—the healing part that people never warn you about. The part where you don't see the person anymore, but every spot you visit feels like a ghost town of things you two used to share, and you still feel a bit hollow in your chest.

I carried that book with me everywhere that summer. The author, Elizabeth Gilbert, pieced me back together. I underlined and circled and wrote in the margins. I felt like I had a traveling companion who understood where I was on the map. And when I finished that book, I knew that was what I wanted to do for the rest of my life. I wanted to write books that made people feel less alone. I wanted to make people feel like they could keep walking forward even when their world had fallen apart. I wanted to be a traveling companion to people who needed to know someone was in their corner.

But I didn't say any of that to the branding coach. My inner critical voice told me I needed a more polished, professional answer. Instead, I told her I was interested in communications, marketing, and millennials. I was interested in how my generation was communicating with one another with all the emerging digital technology, and how we were showing up online. It wasn't a lie, but it also wasn't my heartbeat.

"That's it," she told me. "You're a millennial communications expert."

A *what?*

A millennial communications expert.

What's a millennial communications expert, you ask? I'm not sure I know. But that's what I proceeded to be for the next five years. That's how I branded myself. When anyone asked me who I was or what I did, I told them, "I'm a millennial communications expert."

This was great for me—someone who flounders with small talk—because the questions usually stopped there. People were either impressed or thoroughly confused. Something about those three words, said together, at least made it seem like I had myself together—like I knew what I was talking about and was perhaps even paying my bills.

I put "Millennial Communications Expert" in my email signatures. In my social media bios. On my website. I gave presentations showcasing my expertise. I traveled to colleges and conferences talking about how we, as millennials, could intentionally build a digital footprint for ourselves—one that would say, to anyone who came across our names on Google two hundred years after we died, "I was here. I made something of my life."

I felt a constant pressure to come off as an expert. Always ready to inject commentary. A know-it-all.

———

In the age of social media, it's easy to believe the lie that everyone else has it together and they're all waiting at the finish line for you. At least, that's how I've felt over the years as I've digested books, blogs, podcasts, and social media posts all aimed at bettering one's life.

Everyone on the internet these days seems to be an expert in something. And don't get me wrong: expertise is important. We need experts. We don't want non-experts guiding us on how to treat our health issues, build a house, or cope with depression. But lately, I encounter other people's opinions and beliefs everywhere I turn. Using the internet these days feels like stepping into a networking conference where everyone is trying to mingle and thrust business cards in each other's faces.

And I write these words as someone who has tried to fumble with the business cards and show up as an expert online. I've tried to be that authority. When I started writing those books I dreamed about, I thought I had to have all the answers. Whenever my fingers hit the keys, I felt like I had to know everything since I'd gotten into the habit of looking toward other "experts" to give me all the answers I so desperately wanted: how to live, how to pray, how to grieve, how to parent, how to fix broken things, how to market myself, how to grow.

All the while, life was happening around me. And each life event convinced me that maybe I didn't know as much as I thought.

Maybe you grapple with that same tension—like it's hard even to approach social media because you feel the pressure to always have something to say; you must be constantly qualifying yourself or boosting your credentials. Meanwhile, behind the screen, you have a dying parent. Or you've been trying to make this dream work for the last ten years and you're starting to wonder if it's time to throw in the towel. Or you're beginning to think you might not believe in God anymore and don't know how to say those words out loud. Or you struggled to get out of bed this morning, and no one would expect it (because you seem to keep all the plates spinning), but the pants and the ponytail are a victory for you today.

You may feel like everything you're saying online is a facade, and the last thing you feel like is an expert. I get that. I've been in those shoes before. I used to own several pairs up until recently.

During the time I was moonlighting as a millennial communications expert, I was invited to speak at a Christian women's conference. At first, I was excited and honored, but those feelings quickly fizzled into dread. Pure dread, and a toxic thought spiral: *Who do I think I am? What could I possibly say that would hold value and worth? Are they sure they want me?*

I called my mom to tell her the news, and I'll never forget her response—the first thing she said to me when I told her I'd spend the fall season touring with this group of esteemed women speakers.

"Wow," she said. "Just imagine how much you're going to learn."

She rattled on and on about how much I would learn by sitting at the feet of these seasoned speakers and teachers. "What an opportunity," she kept saying into the phone.

And in that moment, it was like she gave me a free pass to take off the expert cap I was trying so hard to keep secure on my head and instead admit there were many things I didn't know yet. I had so much to learn, and that was absolutely okay.

So that's what I did that entire fall season: I listened to my mom. I carried a notebook with me at all times. I took notes during every talk (even if I had already heard it five or six times). I stayed open to new conversations. I hushed each doubt and fear within me with a simple reminder that I'd been given a chance to learn things I didn't yet know.

———

Just imagine how much you're going to learn.

That perspective has become a driving force in my life, especially as the unplugged hours continue. With each hour of powering down and tuning in to what's right in front of me, my need to show up somewhere else and be some kind of expert is losing its appeal. I'd rather admit there are plenty of things I don't know.

I'd rather be real with you instead of claiming that I have it all together. Sometimes motherhood feels like a mess, and I'm just trying to hold it together and keep my child from becoming a serial killer. I don't know how to keep dirt from getting under my fingernails. I can barely do laundry without everything turning pink or shrinking down to toddler size. I have to constantly recommit myself to acts of self-care, knowing I cannot pour from an empty cup. My faith holds more questions than answers, and I love it that way because I think God loves an honest question. I'm not sure what I want to do when I grow up. And after nine years at my church, I still don't know how to get there without GPS (that might actually be a problem, now that I think of it). But I'm learning all the time that life is infinitely more fun when I no longer believe I have to hold the world together with my opinion, my knowledge, or my commentary.

And here's a side effect I'm experiencing: I'm trying to "fix things" less. I'm becoming confident enough to let go of my need for answers. I'm becoming stable enough to sit with a friend who has been through unspeakable loss and say nothing. I'm wise enough to know that I can't mend everything with my words, that some things can't be fixed on this side of heaven. I'm okay with admitting I don't know. Each admission that I'm a retired know-it-all brings more freedom.

My husband has always been this free; it's one of my favorite things about him. He approaches life with passion and zest. He's eager to learn as much as possible. He reads books just because he finds the topics fascinating, not because they'll propel him further in his finance career. He's gone to extensive lengths to learn how to marinate meat with the right spices and figure out how to cook it in all forms and fashions because it's a pure joy for him to create a meal for others. He reads history books; he watches culinary shows; he's on a quest to read all the words of C. S. Lewis just because he likes how the man thought and wrote. The other day, I caught him watching video after video on oyster farming. He imagined for a day that he could be an oyster farmer, but the beauty is that he moved on to something else the next week. He doesn't feel pressure to have something to show for what he's learning. He's just learning, and he loves it.

So I've tried to adopt some of these ways for myself. I make weekly trips to the library because I remember how much I loved taking frequent trips there as a kid. Rather than scrolling on my phone, I'm reading novels just for the fun of it. I'm baking things because it makes the house feel warmer and more inviting. Seeing the dough rise, like its own little miracle, fills me with more joy than I ever anticipated. I'm taking online courses that have nothing to do with improving myself. I'm reading history books. I'm doing crossword puzzles. I keep notebooks and digital documents all titled the same thing: "I Am Always Learning." I write down a new note whenever I learn a new bit of knowledge about business, life, motherhood, or miscellaneous topics.

When I have no clue how to proceed with something, I jot down a quick note to myself—a little pep talk and a reminder: *It's not on you to know it all. That's not the goal anymore, love. Let yourself off the hook.*

This book's first draft was filled with small notes I wrote to myself at junctures when I wasn't sure how to make the words come together or say the right thing. I'd write myself a note, walk away for a little while, and later return to these thoughtful and curious notes. I was almost shocked to find I was the one who wrote them. Each message was a confirmation: *It's okay not to know yet. One day, you might.*

I spent so much time in my twenties thinking I had to prove myself, almost like I believed if I knew enough, I'd be worthy to take up space. What a small and stifling way to live. Today, I'm a student. I'm liberated by each chance I have to shrug my shoulders and say, "You know what, I don't know. And that's okay. I'm still learning."

The space is wider here. I think I'll stay.

chapter sixteen

ENJOY THE FEAST

Last week, I grabbed dinner with my friend Tory, whom I hadn't seen in a while. When I first moved to Atlanta and knew approximately two people, Tory was there—inviting me to sit at her table, dance at her wedding, and spontaneously grab tacos (always tacos) on a random Tuesday night. When I was painfully single, she and her husband would invite me over to spend time with them, and I'll never forget those days of being included at their table. They'd cook for me. We'd spend hours taking personality tests, psychoanalyzing one another down to the core.

Life led us into different seasons for a while, and we lost touch. It wasn't until recently that we reconnected over a text message.

As I get older, I find that's the beauty of friendships—they sometimes weave in and out of our lives like a needle with thread. You might lose touch with someone for a little while but then life brings them back around unexpectedly. It might be that you need to borrow their view of God when yours feels murky. It could be that they've walked the path you're standing on and they can help you see the light through the trees. For whatever reason, known or unknown, you sometimes find yourself back with someone you never wanted to say goodbye to, and it makes life suddenly more bearable.

My friendship with Tory reignited instantly when I received her text—like an ember that knew it had a spark left in it. And thank God for that ember, because I have so needed her presence in this season of life. Over big plates of tacos and multiple mini bowls of salsa, we lifted our glasses and toasted to reconnecting after several years.

"Confession," she said to me, placing her glass down. "I muted you on social media for a while. I think that's why we weren't connecting as much."

Muting is a soft way of limiting what you see from someone on social media without unfollowing or disconnecting from them entirely.

She went on to tell me about her deep struggles with comparison over the years. She told me how she'd see beautiful things playing out in my life through the screen, and she'd feel a tinge of jealousy. The more I popped up online, the more it triggered a spiral of comparison.

She was keenly aware that the comparison spiral had nothing to do with me, with us, but that there was something more deeply rooted she needed to get to. She muted me to deal with her issues. She faced her feelings. She wrestled with them. And when the time was right, she brought "us" back to the forefront. She protected me from what she was experiencing rather than projecting those feelings onto me. Tears welled in the corners of my eyes as she spoke because I've stood in those shoes of comparison before, and I've felt the pang of jealousy morph into something bigger. I've watched my feelings change toward individuals I genuinely love and admire because of things I saw happening on a screen. We always say comparison is the thief of joy, but it steals so much more than just joy. Left undealt with, comparison robs us of nearly all the abundance life offers. It taints the way we see the world. It changes how we see ourselves. It gets in, and something deep within us starts to rot.

But the truth is, no one stands to win when we're holding measuring sticks. To evaluate ourselves and others based on useless little metrics is to admit that someone will always have to be below us for us to feel secure. That's no way to live. We have bigger things to believe in than that.

I lifted my glass again to propose another toast—to bigger things than measuring sticks.

"Thank you," I said to her when my glass touched hers. "Thank you for muting me."

———

I likely don't have to tell you that social media fuels comparison. We already know this. The only person I know who doesn't compare themselves to anyone else online is my mom, and she's like the love child of Mother Teresa and George Clooney, so there's little hope for us. She still thinks Facebook is a digital house party. If someone is getting too political or unhinged with their conspiracy theories, she goes ahead and unfriends them. It's her way of saying, "You're sucking the fun out of the party—this place is for funny memes and photo dumps."

For the rest of us, it's brutal out there sometimes. Writer Rebecca Webber explains,

Social media is like kerosene poured on the flame of social comparison, dramatically increasing the information about people we're exposed to and forcing our minds to assess. In the past, we absorbed others' triumphs sporadically—the alumni bulletin would report a former classmate having been made partner at the law firm or a neighbor would mention that his kid got into Harvard. Now such news is at our fingertips constantly, updating us about a greater range of people than we previously tracked, and we invite its sepia-filtered jolts of information into our commutes, our moments waiting in line for coffee, even our beds at 2:00 a.m.[1]

Social media can so quickly become a snare. One moment you're scrolling—feeling at least semi-content with yourself—and the next you're filled with intrusive thoughts about your house, education, hair, career path, and the person sleeping next to you.

It's like that game *Minesweeper*—the old desktop game that served no real purpose but to produce intense anxiety in us as we clicked aimlessly around the digital minefield until a bomb blew us up. My hands are sweating just writing about it.

Each piece of content you consume is a click, click, click, until—boom—you're set off. You're irritated and angry. You snap at your partner. You're suddenly exhausted and unmotivated. You lose an hour. You no longer feel good about your progress. It's wild how one small snippet of someone else's day can instantly implode yours.

I'm convinced we could be the happiest people in the world, celebrating our greatest victories in our favorite places on earth, and still, if we left the moment and began to scroll when we didn't plan to—*boom*.

There was a moment a few months into the unplugged hours when I caught myself scrolling after Lane got home. I went upstairs to shower but got distracted. I sat on the bed with two pillows propped up behind me, scrolling. And because I wasn't in the best headspace when I sat down, it wasn't long before the comparison thoughts started shuffling to the forefront of my mind. *You're behind. She's working so much harder than you. You're just sitting here on the phone. They're hanging out—of course they're hanging out. You won't ever be invited into those kinds of spaces.*

What's wild about the scrolling is that we know it's not benefiting us after a certain point, but a lot of us continue anyway. It's there, in that addictive spiral, that we need to remember that our struggle to stop scrolling is no accident. These portals we're holding were designed using the same techniques gambling businesses use to keep their customers placing bets. And according to research, "their methods are so effective they can activate similar mechanisms as cocaine in the brain, creating psychological dependencies."[2]

In the thick of my own doomscrolling, I heard laughter coming through the upstairs bathroom window. It was a late winter day, but spring had decided to show up with a preview of the coming season. Eager to lap up the warmth, I had opened all the windows earlier in the day to let the house breathe.

The laughter got louder and louder. I could make out the voices of my daughter and husband—their delight palpable. I heard the plastic wheels of the brightly colored Cozy Coupe barreling and squeaking through the backyard. I pictured my husband and daughter as I'd seen them a thousand times before— him pushing her through the grass. Her squealing wildly. Both of them yelling, "Faster! Faster!"

And then something weird happened. I felt like I was outside my body, suddenly watching myself from across the room. I saw myself sitting there, mindlessly thumbing through images on my phone, and I could see a dark cloud of comparison and negativity hovering over my head. And then the cloud popped. The pop rang like a warning bell: *Your whole life is downstairs. So what are you doing up here?*

I immediately stood up, powered down the phone, and walked downstairs to join them outside. I found them just as I'd pictured them—careening through the lush green grass together and having the best time. Novalee locked eyes with me instantly as I emerged in the doorframe. She cried out, "Mama! Watch me! Mama! Watch me!"

"I'm watching, baby," I said, taking a deep breath in and sitting on the stairs leading down into the backyard. "I'm watching," I repeated.

The ache of comparison didn't leave my body immediately. I could still feel it buzzing as I sat and watched the two of them, but the more important thing was that I went downstairs to find my life. I often must remind myself that my life is right in front of me, and I'm determined not to miss it.

The thing about unplugging is that the feelings of comparison don't just go away. If anything, the more I power down, the more intensely I feel them, because my spirit is becoming more sensitive to the triggers. But here's the remedy I've been applying since that day: whenever a blast of comparison hits, I get off my phone. That's it. No three strikes—I step away for a little while whenever a tinge of comparison appears in my thinking.

I power down in an aggressive effort to spare my heart. To center my soul. To remind myself, *You are not called to live that person's life—you need to be busy living yours.*

I like social media, but I'm also determined to keep it in its place. If it ever starts to act like the driver, I step away.

Before anything else, I step away from the phone and consciously step back into my life. Because that's the very thing that withers when I don't stop scrolling—the life I could have had if I decided to stop measuring and start investing instead. I pick up a book I've longed to read, or I make a cup of tea. I write out a card for a friend or I cuddle my daughter. I do something physical to break me away from the digital and to remind myself: *This moment needs your attention.*

———

I don't want you to miss your life, friend.

I know many voices circulate in your head. I know one voice often tries to get you to believe there is no room for you here. That all you'll ever do in this one life you have is watch other people get the things you want. The marriage you wanted. The positive or negative test you wanted. The friendships you wanted. The book deal you wanted. The awards and the titles and the celebration you wanted.

And I wouldn't want to discount all you feel, not for one second. Your feelings are real, and they're hard. And some days, no amount of "celebrate others" or "just stop comparing" will fix the ache you feel, the pain that makes you feel like you've been forgotten, you're not good enough, and it will never be your season. I'm sorry. Those are lies. But lies can feel believable and hurt us just the same.

I keep thinking about the parable of the prodigal son as I write these words to you. It's a story about a father and two brothers. The younger brother leaves home, takes his inheritance, wastes it, and crawls back to the estate. He expects to live as a servant, indebted for his mistakes, from now on. But upon his return, the father runs to him. He celebrates him. He plans an enormous banquet.

The older brother—who gives off major perfectionist and goody-two-shoes vibes—comes home from a long day of working in the fields to find this big banquet happening inside for a brother

115

he hasn't seen in a long time. The older brother is angry and refuses to go in because he feels like his brother doesn't deserve the feast spread out for him. And, also, because his brother is getting what he himself so desperately wanted.

The father steps out of the banquet to tell the older brother the truth: he's always had this kind of access, not just to a feast but to a whole abundant life. It was all right there in front of him, but he never had the eyes to see it.

That's how the story ends—with the two of them standing outside the banquet. No big resolution. No dramatic reunion. We never find out if the older brother went inside and enjoyed the feast.

I used to feel like the older brother in that story. I identified with him so much. I was consumed by fear that there wouldn't be enough for me, and I missed many feasts as a result. But it hit me one day as I was reading a commentary on the parable: *I don't want to be the older brother in the story anymore. I don't want to be the one who misses the feast because of my own feelings.*

If a feast is happening inside, I want to be present for it. If goodness is happening for someone else, I want to bring flowers. If someone posts a beautiful piece of news online, I want to be in the comments section, lifting them up with enthusiasm and genuine joy. I don't want to look back and realize that my life could have been an abundant feast—the kind I was searching for the whole time—if I'd just stepped into the banquet hall instead of spectating from the window outside.

Staying a spectator is how you miss a life. I'm afraid it's that weighty. You cannot possibly be in your life completely if you're always looking to the right and left of you or focusing all your energy on how you measure up against those around you. You can't possibly see the sacred hiding in plain sight when your gaze is fixed on fear and lack.

But it's not too late. There are things ahead in your story that only you are wired for. Certain bits and pieces that will only ever be yours. Moments when it will be beyond clear that this isn't an accident—that there's nothing accidental about you and this path you're on. There will be moments of feasting when you look up, look around, and say, "Wow. Just wow."

When those moments arrive, I hope you're *in* them. I hope you embody them fully. More than anything, I hope you enjoy the feast.

chapter seventeen

LONGER TABLES

Last year, I attended a retreat in the mountains of Utah and found myself sitting at one of the most beautiful tables I've ever been seated at. Greenery and tea candles covered the long rustic wooden tables throughout the room. There were carefully constructed name cards at every plate. The lights were dimmed, filling the room with a striking ambiance that stirred peace in my spirit after a long season of scrambling.

I sat beside women who, though I'd just met them a few hours earlier, inspired me to my core. They all came from different industries and different junctures in their careers and lives, but there were common threads: they were kind and thoughtful, bold and radiant.

The evening was intentionally crafted to make space for deep conversations—the kind we're often too busy to have in everyday life. We broke up the long tables into more intimate pockets of three and four. For each course of the meal, a different question was posed. The questions took us on a journey of getting to know one another—layer beneath layer. There was a feeling of safety within those small circles as we dared to share the parts of our stories we were afraid to utter.

Something about this environment revitalized my spirit. I felt like I was coming home to myself after a long trip away. So many things over the previous few years had left me feeling less confident in my spirit, and I'd slowly been trying to heal and repair. My experience that night felt like restoring a piece of me I'd been missing. Like remembering who I was.

At one point in the evening, the chef who'd carefully planned every detail of the menu stood up to talk about the preparation of the meals and the experience we were enjoying. He acknowledged the palpable spirit of competition within the restaurant industry. It's a cutthroat field that demands a relentless pace. He was on a mission to transform the vicious cycle of entitlement in the hospitality industry into a virtuous cycle of generosity. His whole demeanor was poised and calm. His delivery was so thoughtful and articulate that I couldn't help but take my phone out from beneath the table and start taking notes.

"Hosting has nothing to do with the 'looks of things,'" he said. "Hospitality is the art of making people feel valued. We live in a world where everyone is so consumed with showing off their trophies. But true hospitality isn't about showing your trophies; it's about showing your cards."

That's when something in my brain snapped to attention. I'd always carried a surface-level definition of hospitality. I thought it was all about table settings, finger sandwiches, and remembering to leave clean towels in the bathroom for overnight guests.

I thought back to all the instances when I'd hosted others in the past, and how I'd been consumed with ensuring every detail of the environment was perfect. I never realized I was making the whole thing about me. I was waiting for compliments or to be showered with praise. I was waiting to be seen rather than using the opportunity to see others fully. I could have focused all my attention to detail on making other people feel valued as they walked through the door. That would have changed everything—for them and for me.

In our culture, we can naturally get swept up in the aesthetic of things. The constant visuals of other people's lives have made us more hesitant to open our doors to one another. Now there are "looks" to things. There are matching dishes. There are charcuterie boards that take up the whole table. There are Pinterest boards for how our homes should look and feel. And so we keep our doors shut until we think we have the perfect visual to present to others. We miss the opportunity to say, "Hey, come over tonight. The floors aren't swept, and stuff is all over the counters, but who cares? Just come."

These opportunities seem to happen less and less as life becomes more digitally charged, but I think we can be the people who create more of them. We can learn to swing the door wide open. To do that, we must be willing to lay down our standards of perfection. We have to forsake the aesthetic for what actually matters: the chance to get better at showing our love and kindness to one another. If we think the floor must always be swept and the laundry needs to be put away, that our messes must be handled and our lives kept tidy before we can invite others in, I'm afraid we'll never get there.

We experienced this just the other night. The house was a mess, and we were getting ready to go to bed when friends called and said their power was out. Could they come and spend the night in our guest room? At first I groaned, mentally critiquing the unkempt space around me, and then I remembered the chef's words that I now carry in my back pocket: "It's not about the looks of things."

We said, "Yes, come." We left the toys all over the floor. Our friends helped with bedtime. We prepared them leftovers and shared some chocolate cake we had in the fridge. We stayed up late talking, wrapped in blankets on the couch. We didn't look at our phones once.

The other night, a friend of mine hosted a "color party"— one of those parties where you're assigned a color, and you bring food and drink matching that color. I was hesitant to go because staying home and watching reality TV felt easier than putting on my pants and talking to other humans. Not to mention, my color was brown.

As it turns out, all the best and most nostalgic foods are brown. I piled a wooden cheese board full of Hershey bars, semi-sweet chocolate chips, pretzels, Little Debbie Swiss Rolls, and bacon-wrapped dates.

I walked into my friend's house with my brown board. The counters and tables were already filled with brightly colored displays. Someone carried in a yellow board made up of buttery popcorn, pineapple, and cheese. Another girl brought a collection of strawberry shortcakes. A friend appeared in the doorway with a bag of blue corn chips and a bottle of champagne with a blue label.

The house was filled. Women sat around the table. Women spilled into the patio space outside. Women stood casually by the counter and caught up. Many of the women didn't know one another. It was an open invitation—if you knew someone who needed friends in the city, you should bring them or tell them to come. I stood back and watched the evening unfold as women met, mingled, and broke into smaller groups or sat in big circles throughout the house. It quickly became clear to me that this evening had nothing to do with colors.

This was belonging. This was Hospitality 101. *Make room. Extend the invite. Open the door.*

My favorite moment of the night: I watched a girl come into the house with a sleeve of Oreos. No fancy board. No big display. She placed them on the kitchen island and walked into the circle of women sitting outside. She'd gotten the memo: Bring yourself. You don't need to try to impress anyone. Just come—there's space for you here.

I called my friend the following day to thank her for making the space. Over the phone, she told me the origin story of these evenings. She had walked through a season in which she felt uninvited. She would often see people on social media hanging out without her. Instead of being the outsider, she decided she would be the inviter. She built a longer table.

I think this shift is for all of us. We're all called to be practitioners of hospitality—to be the inviters. We can be the builders of the longer tables.

We don't need to wait on extravagance. The people in our lives don't need fancy tables and string lights. They don't need big displays or a Michelin restaurant experience. That's not why real humans show up at the door anyway. After years of isolation, I think most people are just hungry to return to a space of communion with one another—where we can remember an essential truth: "You belong here. Many spaces in life might make you wonder if you belong anywhere, but you belong here. And you can stay as long as you'd like." I can't think of a more worthy calling than to be the ones who make spaces where others can find safety and restoration when they're at the end of themselves.

Perhaps genuine hospitality is a frozen pizza in the oven, a messy house, and a question asked after a long day: *How are you? Really? Tell me the truth.*

We can take the pressure off.

Don't even worry about clearing the countertops.

Show your cards, not your trophies.

chapter eighteen

STATUS UPDATES

When I was a senior in college, a young woman from our campus ministry office approached me one afternoon with what she told me was a message from God.

She was a fellow student, but up until now, I hadn't had many interactions with her—and that's likely because she was getting messages from God. Her spiritual maturity level seemed way out of my league, and I felt like I could never approach her. My spiritual depth at the time was equivalent to a plastic kiddie pool.

I was sitting in the common space outside the campus ministry office when she walked up to me and handed me a small scrap of paper, folded like one of those compact notes you would pass beneath the desks to your middle school crush.

"God has really been pressing this word into my heart for you," she said. "This word will mean a lot for you in the coming years." Without waiting for my reaction, she turned around and walked back into the campus ministry office.

I remember being a skeptic. I wasn't fully convinced God could even speak, never mind speak to someone else about me.

I unfolded the small scrap of paper to behold one word delicately etched in pencil: *Vulnerability.*

At the time, the word *vulnerability* wasn't the buzzword it is now. This was the pre–Brené Brown era. So I looked up the definition on one of the library computers: "capable of being physically or emotionally wounded; open to attack or damage."[1] I was part horrified and part deeply intrigued by this definition. I tucked the message away in my shoulder bag, curious about the interaction with the young woman.

I've remembered that tiny scrap of paper many times over the last decade as I've forged a life of writing things on the internet. More and more, that was the comment I heard from other people as I pressed publish online: "You're so vulnerable with your words."

And while I do believe that sharing parts of your life online involves vulnerability, I think that for me it eventually morphed into oversharing.

Unhealthy behaviors emerged when I started viewing my whole life as a piece of content and processing it all online. Nothing was off-limits. The lines between public and private had blurred, and I had no idea which words should be kept to myself. Or my smallest inner circle. Or the skinny lines in my diary. Or my prayer closet. Or any other places that weren't on the internet where other people could consume them.

One guy I was involved with even told me, as we were breaking things off, "I don't want to become a piece of content for you." I wrote about him the next day.

Sadly, I saw every conversation as a piece of content. I saw every prayer tucked away in a journal as something to be shared with someone else. I stopped experiencing and processing the realities of my life because I was too busy making sure my life looked good to other people online.

My wake-up call came at a speaking engagement. It was one of my first since I'd army-crawled my way out of a depression that threatened to take my life. I was doing so much better. I was healthier, more connected to myself and others, and more myself than I'd ever felt. A woman approached me from the back of the room and commented on how differently I'd spoken that day versus the year before. She told me she'd followed my journey on the internet, and she could tell I was going through something.

"My therapist and I always talk about you," she told me. "She asked me if I thought you were depressed. And we both agreed that you were."

It felt like an out-of-body experience to know someone was spending their precious therapy dollars talking about me and analyzing my mental health. I wanted to write her a refund check for all that therapy. But I couldn't fault her for being concerned—I'd opened up my world to so many people, and she'd gotten used to knowing all the parts of me that I shared online. And I could tell she genuinely cared about me. But I knew in that moment that something needed to shift. I needed to step away from oversharing and step into my life instead. Better yet, I needed to allow close friends and family to step into my real life with me—daring me to face it and not move away—when I couldn't tie it with a bow or make it tweetable.

———

When we're online, sharing and consuming all the time, it's easy for the lines between reality and curation to blur. The more we congregate in online spaces where we can edit and filter, the more we start to favor our fragmented displays. However, Vivek

H. Murthy explains that the way we present ourselves online—as prettier, happier, braver, or more successful versions of ourselves—only fuels our social withdrawal. "These poses . . . may let us pretend that we're more accepted, but the pretense only intensifies our loneliness."[2]

Sherry Turkle, one of the most prominent voices in the conversation about the way technology is affecting our modern relationships, agrees. She claims,

> Human relationships are rich and they're messy and they're demanding. And we clean them up with technology. And when we do, one of the things that can happen is that we sacrifice conversation for mere connection. We shortchange ourselves. And over time, we seem to forget this, or we seem to stop caring.[3]

If I'm being completely honest, isolation often feels more manageable than connection. But I'm learning that even if it feels weak to admit it, you and I need people. Some truths are hard, and this is just one of them: We need actual, real-life people to see our whole selves—not just the fragments we pick and choose to display. We need to be able to process the events of our lives in spaces where moving forward and growing feels suddenly possible because we're not doing it alone. We need people who can guide us, teach us, and share their insights with us. And we need people who can challenge the narratives we've believed for so long— stories we've convinced ourselves must be the truth even though they fill us with shame or fear or doubt. We need people who give us one look, open their arms, and say, "Your truth doesn't scare me. Tell me all of it."

I've learned through the unplugged hours that honesty and vulnerability are two wildly different things, though we often think of them as being the same. Honesty is like opening the front door and allowing others to peek in. Many of us practice honesty on social media, sharing parts of our hearts, our bigger fears, our shortcomings, and our crazy dreams. But vulnerability goes a step further. It's not just opening the door so someone can see our mess; it's inviting them inside to be in it with us. It's saying, "Yes, you can come in," when we've got nothing to offer them. It's being willing to be helped when we don't feel shiny, special, or worthwhile. I don't think we can practice that kind of vulnerability with people on the internet. That kind of vulnerability requires us to take off the mask and remove the filter. That kind of vulnerability often requires us to be face-to-face.

I'm not saying this will be easy. Or that it will be perfect. When we open ourselves up to needing others, we expose ourselves to the potential of getting hurt. It would be easier to stay behind our screens—to keep our interactions surface-level. But I'm afraid our relationships don't produce much fruit when they stay surface-level—they need to establish deep roots, pushing through dark places and reaching for light, before they yield the fullness and richness of connection that we long for. But here's something to look forward to: The most uncomfortable steps of vulnerability yield some of the sweetest harvests after we get gutsy enough to take them.

———

During my unplugged year, I went on a weekend trip to a lake house with my girlfriends. They're a core group of women I've known since I moved from Connecticut to Atlanta nearly ten years ago. We've seen each other through engagements, marriages, pregnancies, babies, losses, moves, transitions, and potty training—all of it. Sometimes I marvel at our early stages of friendship, how a group of girls getting together to watch *The Bachelor* and paint our nails on Monday nights has morphed into this.

The weekend was entirely unplugged because of the lack of cell service, but everyone seemed to lean in. We hiked and caught up on sleep. We made charcuterie boards throughout the day, read by the lake, and took turns updating one another on life. Husbands. Kids. Jobs. Transitions. I love that the updating didn't happen all at once, right at the beginning of the weekend. Hour by hour, our status updates—the unedited versions—slowly came to the forefront between more coffee being brewed, more cheese being set out, and more cups being filled.

I've learned that unplugged atmospheres do something to us. They possess their own kind of rare magic—they open up space for honesty, depth, and revelations that turn into breakthroughs. It's a type of refreshment we rarely get when our devices are all plugged in and pinging. These environments free us and sometimes even show us where we're trapped but didn't know it.

Before heading out to dinner that Saturday, we sat on the back porch and the conversation turned to me. I'd been doing everything possible to avoid this moment, sugarcoating all my stories so I didn't have to be vulnerable. But somehow we got onto the topic of the body image issues I've struggled with for years.

Even now, sharing this struggle with you and being so open and exposed on the page makes me want to recoil. This has been a battle for me since I was in my early twenties, and the fact that I've been dealing with it for so long seems to add an extra layer of shame. I want it to be over by now. I can't look back on certain seasons in my life and not see body image issues looming in every mirror selfie, every pair of pants, every group picture. I spent years being hard on my body. I restricted myself in so many ways, in so many efforts to get smaller.

Like a weed that should have been pulled when it first

sprouted up, the body image issues spread into almost every other area of my life. When you hate what you see in the mirror, there's no way it doesn't impact your marriage, your children, or your vocation. It affects everything, even if it's just running on autopilot in the background.

"When do you think it all ramped up?" one friend asked.

"Sometime around the time Lane and I got married," I said.

The circle gave me a reassuring look because they knew exactly what else had happened during that time frame—I'd just never put the pieces together before. The start of my marriage was also the end of one of my closest friendships—a friendship I thought I'd have forever. And because we rarely talk about the reality of sometimes losing friends as we grow older, I didn't know how to deal with the grief. I was consumed by it. I stopped recognizing myself. I started harboring anger and a deep self-loathing. I stopped wanting to take up space. I stopped going to certain places out of fear that I might bump into my former friend. I was shrinking back in every imaginable facet of my life.

I started buying clothes several sizes too big to cover up the shame rattling inside me. I don't know why, but I told myself a story that said my friend deserved to thrive—and I didn't.

But here, in this circle of people who were for me, I admitted it all for the first time. I acknowledged that I had never felt quite like myself since the day that friendship ended, and I had been struggling to move on ever since. I let these women speak life into me. I let them love me and see me without any edits or filters. And it felt like a little voice within me was whispering, *You can start to heal now. It's okay. You're safe here. You're fully seen.*

Author Donald Miller once wrote,

When the story of earth is told, all that will be remembered is the truth we exchanged. The vulnerable moments. The terrifying risk of love and the care we took to cultivate it. And all the rest, the distracting noises of insecurity and the flattery and the flashbulbs will flicker out like a turned-off television.[4]

That's exactly how that moment of admission felt for me. It felt like throwing open a door that had been shut for years and saying, "Here, come in, I've got nothing to offer, no answers or direction just yet, but I don't want to be in here alone anymore."

My friend Christina asked what the first step toward healing might be. I told the little circle of women a silly story about how I'd gone into a clothing store the week before and bought a crisp white button-up shirt that made me feel good when I put it on. I thought, *This is what I would wear if I were confident. If I were truly myself.*

As I held that shirt in the store, it didn't feel like just a piece of clothing; it felt strangely like a fresh start.

Getting ready for dinner that night, I pulled out the white shirt and held it up. I tried to shove it back into the suitcase and go for something more familiar.

My friend Christina looked at me and said, in the gentlest voice, "The white button-up." And I instantly knew what she was really saying to me: *You're not going back into hiding. The mask is off now. Take one tiny step out into the light.*

Nothing magically shifted when I wore that white shirt and a pair of jeans. I felt uncomfortable that whole night at dinner, and I honestly wanted to go home, change out of the shirt, and put on something else. But perhaps the more important part is that I didn't. I didn't shrink. I didn't go back into hiding. I pushed through the uncomfortable, and I stayed there, willing myself to take up space.

chapter nineteen

WONDER

My two-year-old walks around this world in perpetual awe and wonder.

I often have to remind myself that this is her first rodeo. She's encountering the world for the first time—she can't help but take in all the sights, scents, and sounds. She doesn't have twenty years under her belt. To her, everything is brand new, and it's all worth noticing.

Chocolate is irresistible because she's never tasted something so sweet. Automobiles are amazing to her because how could they not be? Fireflies flickering through the grass at night are something near holy to behold. At one point or another, these things amazed me too.

Lane and I joke that her perpetual question, "What's that?," is the chorus of our lives. She notices everything. The makes and models of cars beside us. The way the car engine accelerates when the light turns green. She marvels at the tones and melodies of a worship song even if she doesn't fully grasp the depth of the words just yet. She points out willowy trees and airplanes flying overhead with an excitement I lost a long time ago.

She zooms her baby stroller around the backyard, occasionally stopping to say, "Hear dat, Mama?"

I crane my neck and turn my ears toward the sound. I wait until I hear the subtle rumblings of a train off in the distance. Until she started pointing out the sound several times a day, I had never heard a train or noticed an airplane flying overhead. I've lived in this house for five years.

She looks at me as if thinking, *How are you not more amazed by all of this?*

I wish I could paint myself as an especially present mom who notices every sight, sound, and color. But I'm often hurried or preoccupied. I'm telling her to speed up when we're walking. She's staring out the window at all the sights, and I'm listening to yet another podcast on simplifying my life through a complex organizational system.

Lately I've wondered, *How many times have I unintentionally snuffed out her sense of wonder?* Out of hurriedness. Out of impatience. Out of something "more important" on my phone.

Since the start of the unplugged hours, I've been thinking more about the idea of wonder and wondering when it slipped out the back door for me. Maybe it was entering the perils of adulthood. Maybe it was student loans. Maybe internet trolls are responsible, or perhaps tax season is the culprit. I even searched the internet for answers about what makes us lose our sense of awe and wonder. I read many scholarly articles on the topic, but I think Reddit user "Tonk" said it better than anyone else: "Life is very, very hard. It takes practice to maintain the wonder."[1]

I recently read a series of studies from researchers at the University of California, Berkeley, who set out to explore the effects awe and wonder have on us. In one study, they took two groups of people and exposed one group to an experience that induced a sense of wonder. The other group was not given the same experience. Both groups were asked to draw a picture of themselves afterward. Those who had experienced wonder drew themselves significantly smaller in the picture than those who had not experienced wonder.[2]

We've all encountered that sense of awe-inspired smallness— whether we've lain beneath a starry sky or sat and watched ocean waves crashing toward the shore. It's not the kind of smallness that makes us feel insignificant or worthless; it's a deep sense that we're enveloped in something far more vast and intricate than we ever imagined.

This phenomenon has an ironic name. It's called "unselfing." Unselfing is what happens when we turn away from ourselves and look outward. Our bodies react in pretty miraculous ways. Our nervous system calms down, and there's a release of oxytocin— the chemical that stirs up feelings of love and trust within us. We feel more connected to others. Our worries hush. Our mental chatter grows quieter. We become less self-absorbed. The part of the cerebral cortex that regulates how we see ourselves deactivates. Our negative self-talk takes a break. Our connections grow deeper, and our relationships get stronger.

"When our sense of self decreases, you might think that we would feel that what we are doing is less important or less significant. But there's a paradox: as our sense of self diminishes, we feel more significant in other ways," write Jake Eagle and Michael Amster, authors of *The Power of Awe*. "We feel momentum, and we are moving with the momentum. We aren't just on our own anymore. Something is happening, and we are part of it. We feel connected."[3]

It's evident our bodies and minds respond favorably to experiences of awe and wonder, but it's much harder to experience them when our heads are always down.

———

I've created a breath prayer for this unplugged season of life.

Breath prayers are exactly what they sound like: short prayers you can utter in one breath. Sometimes talking to God is hard and you feel like you don't have the right words. That's when a breath prayer comes in handy. You can say it on repeat. When you're driving in the car. When you're fighting off anxiety. When you're trying to learn to be more present in the moment. Eventually, you start to believe the words and the prayer starts to feel as familiar as breathing. Hence the breath part.

My faith is a patchwork of breath prayers that have moved me from one season into the next. So in this season, I pray, *God, restore my sense of wonder.*

A longer version of the prayer goes like this:

Dear God, restore my sense of wonder.

Because my wonder ran off and I didn't bother to chase it. Because I stopped noticing things that were always meant to be miraculous. Like the people interacting in coffee shops. Like how puddles form on the roadside after we've said too many times, "We needed that rain."

Restore my wonder, because I go through the motions so often and take this whole "being alive, walking around this planet with oxygen in my lungs" gig for granted.

My wonder has flatlined. It has taken far too much

134

NyQuil and drifted into a deep sleep. Please wake up my wonder. Use force if necessary.
Amen.

This is one of those prayers I recommend putting on a Post-it note and sticking to your dashboard or someplace where you'll see it repeatedly. Sometimes we must remind ourselves of the prayers we prayed yesterday, the day before, or a year ago. Sometimes we're already living in the answers without even recognizing it.

The other night, my husband and I met old friends for a concert at a tiny attic venue in Atlanta. The words *concert* and *me* are not typically two words you'd place together, but my friend assured me I'd like the music because the artist is a storyteller and a hymn writer.

We shared a meal at one of the small tables scattered throughout the venue and caught up on life. The show began. A small-framed woman emerged on the stage, hoisted a guitar to her hip, and began to play. Throughout the evening, I kept looking around the small space at everyone enjoying themselves. Laughing. Singing along. It was my first experience being in a crowd since the pandemic. I marveled at how we once thought we might never have these gatherings again. It meant more this time around.

Toward the end of the concert, the artist began strumming an old hymn. We all knew it. In this little attic space, we all sat close and sang together. And for just a few moments, it felt like the tiniest taste of heaven on earth. It was as if the wonder I'd been missing pulled up a chair beside me and nudged me, saying, *Stick close. Your tired spirit will be restored. This is why we do this whole "humanity" thing. For moments like this.*

————————

I have to believe wonder can be restored at any age—that we're never too old to enter into a space of awe. To get there, we must be willing to not know everything. To loosen our grip. To hold off on the Google search and just sit with the unknown for a little while longer. Maybe there's an answer, maybe there isn't. Maybe the answer doesn't matter right now. Maybe we're allowed to wonder our way through it.

If we're paying attention, wonder can reveal itself almost anywhere. I think the biggest lie we've believed about wonder is that we must leave our daily lives to find it. We're convinced we need plane tickets and vacations to reach the place where wonder lives, but wonder has always been right here in front of us. Hiding in plain sight.

It starts with noticing. Looking up just a little bit more. Taking time to watch the spider spinning a web in the corner. Bringing the morning coffee outside so you can breathe in the summer air for a few minutes.

These are the things that restore us to smallness. They remind us, *Hey, you're not holding the world together. You're part of a bigger story. That thing you're worried about, I know it feels so big, but resize yourself. This, too, shall pass. Keep looking up.*

Wonder even makes house calls.

The other day, Novalee woke up from her nap to the sight of a glowering thunderstorm outside. She was completely captivated. She wanted to stand at my office window and watch. I can typically marvel at a rainstorm for about thirty seconds before I'm on to the next thing. But Lane has a pace that perfectly matches Novalee's, and so he reveled in her awe.

"Let's go downstairs and get a better view," he told her. He scooped her up and down they went to the front door. He swung the door wide open. The two of them plopped down right there in the doorway. He held her in his lap, and they counted the gaps between the lightning flashes and the thunderclaps.

I had followed behind and paused, my eyes tracing the outline of their figures in the doorway. Snap. A mental photograph before sitting down with them in the doorway to watch. The thunder and lightning riffed off one another, hamming up their performance like they'd been waiting years to show off their talents. The three of us huddled closer. We said nothing for what felt like a really long time. We were a captive audience.

part three

RHYTHMS

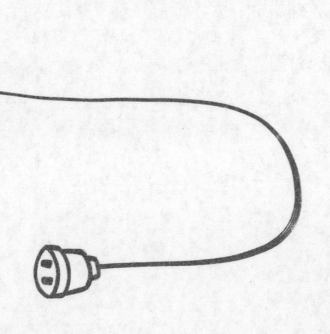

chapter twenty

DELIVERED QUIETLY

Lane and I celebrated our most recent wedding anniversary with an empty calendar and a quiet weekend at home.

For many years, we always took the opportunity to get out of town, but in recent years we've reveled in the chance to stay in. We drop off Novalee with the grandparents and enjoy date nights at old favorites. We take our books and Bibles to the local coffee shop, sip our coffee slowly, and no one has a meltdown over the red velvet donuts in the glass case by the register. We remind ourselves of the life we're building together. We take inventory of the past twelve months. We re-up our subscription for another year.

This year, we went to the movies and got popcorn with fake butter. We ate biscuits and played rummy while sharing champagne at the spot where we had our first date. On Sunday, we slept in, opted for the later service at church, and strolled into the sanctuary feeling rested and young.

In the dark of the church sanctuary, I settled in to worship. Admittedly, it typically takes me a long time to get to a place where I even want to worship. I must fend off a million little to-dos in my mind first. It's as if all the small tasks—remembering that email, buying more laundry detergent—are perched at the edge of my mind, waiting for a quiet moment to dart at me with a vengeance like those terrifying flying monkeys from Oz.

That morning, though, my mind felt free. A posture of worship came easily. My friend sang from the center of the stage. She does this nearly every Sunday, but something about this time felt different. She was glowing. I had an overwhelming sensation in my spirit that said: *Reach out to her. Tell her she's glowing. Tell her you're proud. Tell her God is proud.* There was an urgency to the sensation that felt almost tangible.

Over the years, I've become increasingly familiar with these sensations. I call them nudges. Holy nudges I can't ignore. And I've noticed that the nudges became stronger and more frequent once I started powering down.

I think these nudges are happening all the time; they hover in the background of our lives, waiting for us to listen for them. The nudges, I've discovered, won't compete with the layers of noise we add to our lives. Either we learn to tune out some of the noise so we can hear them, or we miss the nudges altogether. But, wow, when we move toward a life where we can notice the nudges, everything about the landscape we're walking in starts to change.

A couple of notes about nudges: They don't really care about your comfort zone. In many cases, they will ask you to step out of it. It might be a nudge to call an old friend. A nudge to send a text. A nudge to ask someone if you can bring them dinner tonight.

I believe these nudges come from the Holy Spirit. Early Celtic Christians referred to the Holy Spirit as the "wild goose." They believed the Holy Spirit could be as gentle as a dove yet also as wild as a rambunctious goose—prompting us to step into unpredictable territories, into new adventures.

That's another thing about nudges—they might feel crazy or out of left field, like that wild, rambunctious goose. You might not want to follow through. You might try to back out or get God to send in a sub. You might feel unhinged when you follow through. But I urge you to follow through anyway. And here's why: We don't always know what others are walking through. Or we might feel certain we do know but actually have no idea. We humans are crafty when it comes to hiding our pain, insecurities, losses, and struggles. What we project to the outside world is typically just one dimension of our story.

You might feel ridiculous asking a friend if you can bring them dinner, and yet they might be at the end of their rope—drowning in toddlers—with nothing left in the fridge.

You might feel like a loon texting your most confident friend with a message of encouragement, and yet they might be putting up the biggest front of their life, doing everything they can to keep from falling apart.

You might feel like your words don't matter, that no one will read them, so why even try to put them out into the world one more time? And yet someone else might come across those words you wrote today and see them as the sign they've been praying for.

———

I'm embarrassed to admit I didn't follow through on the nudge I felt in church that day. Life flooded back in the second we exited those church doors. The churning engine of "have-tos" and "must-dos" started again. *Ping, ping, ping*—the notifications rolled in.

We need paper towels.

Your prescription is ready.

That bill is overdue.

Don't forget that meeting on the calendar. Monday at 2:00 p.m.

That email needs a response.

Before long, there were days between that nudge and me. The urgency of the moment faded. Life lurched forward.

———

I'm fighting notification fatigue.

It's an actual thing—a reality that emerged out of the pandemic. Somewhere in that stressful, strange season, many of us adopted an "always on" mentality that slowly eroded our spirits. Now that we're out of the pandemic, I think we're still struggling to lay down some of the habits we picked up during that season. We packed our mental backpacks until they were far too heavy and full, and now we don't know where to lay down the stuff. We don't know how to return to a place where our ears aren't buzzing and our minds aren't always overstimulated.

As I've read articles about notification fatigue, one detail stands out to me: *The more notifications that roll in, the more we're desensitized to the urgent ones—the ones that actually matter.*

Things that could easily be handled tomorrow masquerade as urgent requests. There's an expectation that we will stay on at all times. There's an anxiety that floats in the air when we don't respond to a text within a few minutes. We're all contributing to this culture, but I think some of us are just waiting to be let off the hook on which we've hung ourselves.

I'll be honest, my capacity for "always being on" has diminished. At first, I was concerned by this, but now I think it's one of the blessed side effects of the unplugged hours. I have less capacity for notifications that show up on screens. Less capacity for stimulation, for too many tasks at once, for text threads that go back and forth all day, and for inboxes that will never reach zero anyway.

But I have more capacity in other areas. More capacity for togetherness. For connecting. For deep breathing. For writing letters and reading novels. For waiting—on seasons to change, the water to boil, and answers to come. Life is better overall, but this has come at the cost of my spirit feeling more sensitive to the overload I once allowed into every area of my life.

When I find that familiar sense of anxiety rising in my chest from the pressure to be "on," I intentionally step into a new rhythm I've created. I shut down technology (if I can), put on my sneakers, and walk around the block. I come back to my desk, and I write out a list. I call it my "What's Not Working" list, and it's exactly what it sounds like. I do a quick yet thoughtful examination of what's not working. Most items on the list point back to the same conclusion—I am, once again, too plugged in. Somewhere, when I wasn't being mindful of it, an old habit weaseled its way back into the folds of my life. I am either scrolling first thing in the morning or checking emails late into the night, building mental task lists in the dark. It might be that I haven't powered down in a few days and I need to get back to tracking the hours. Whatever it is, I write down what's not working with no shame or judgment. I get it all out on paper because someone once told me that when we put our thoughts on paper, they become malleable. We can start to address and change them from there.

Once the list is made, I do my best to pair each "What's Not Working" list item with a small solution. I mean, really small. Big shifts might fill me with gusto in the moment, but often they're not sustainable. From there, I start making tiny shifts and planting new rhythms that take the overwhelm off my shoulders. I recommit myself to getting back in working order.

———

The Thursday morning after I had the nudge in church, I sit in a corner at Starbucks and craft a text to my friend who'd led worship on Sunday. I trust my lateness just might be God's perfect timing. Though it took me a few days to send the text, something shifted in my spirit immediately after hitting send.

I open up my notebook and draft my "What's Not Working" list. I connect the dots. I plot my little rhythms. Over the next few days, I make a clean sweep of things. I delete apps I'm not using. I delete apps I am using that add nothing to my life. I clear out text threads. I respond to people I've left hanging. I turn off almost all notifications.

————

While writing these words to you, I've had moments when I wish I could throw aside the yellow legal pad and close the computer, when it would feel better to simply invite you into my corner of the world. This is one of those moments.

You could show up at my door. Just bring yourself. I'd pour you a fresh cup of coffee and show you the place in the fridge where all the creamers congregate. You'd pick a spot on the big couch, and I'd hand you a blanket. I'd bump up the thermostat a few degrees and turn on the gas fireplace.

And then, when we're all settled in, I'd lean in and ask you some honest questions:

Are you overloaded by all the noise? Do you feel it too?
Do you feel this same alarming pressure to always be "on"?

It's okay to feel overloaded. That's what I would tell you. You can start un-overloading. That's not a selfish act. You can decide the system is broken. You can decide with deep clarity that you don't want to get to the end of your life and realize you paid attention to the wrong notifications. You don't want to get to the end and learn you could have heard so much more from God, but instead you flung your attention at dopamine-inducing headlines and false alarms.

You can commit yourself to turning off notifications that don't matter. You can build better boundaries—ones that protect your family, your peace, and your mental health. It won't be a one-and-done thing. You'll have to do it over and over again. Different seasons will call for different boundaries. But you're capable of doing the hard work. Other parts of your life will flourish because of the boundaries you're drawing and the priorities you're protecting.

You can carve out space for what actually matters—for the nudges that are most often delivered quietly.

chapter twenty-one

RABBI

We are a follower-obsessed culture—that much is clear. We follow news headlines. We follow trends. We follow stocks in the market and the relationships of celebrities. We follow diets and we follow advice. We follow spiritual leaders and TV shows, weather reports and sports teams.

When Twitter and Tumblr emerged at the forefront of the social media landscape in the early 2000s, the notion of "friends" we'd grown used to through MySpace and Facebook shifted into "followers."[1] The world changed, and we changed with it. Suddenly, we became interested and engaged observers of the lives of people we didn't know. We started consuming bits and pieces of one another's lives—thoughts and opinions, workout routines and organization hacks.

So many of us wake up following. We go to bed following. I know the morning routines of some people I've never met (which is both strange and sort of fun). I follow some cold cases so closely that being an armchair detective feels like it's a second job. I follow parenting accounts and cooking accounts, and there are some people I genuinely love watching through a screen even though I've never met them in real life.

But more and more, I've started to wonder if Jesus knew when he declared, "Come and follow me," that we would be here, two thousand years later, deeply obsessed with following one another so closely.

———

When the unplugged hours began, my faith was going through what I'd call a sifting process. As a writer, I care deeply about the words I use to describe things, and "sifting" feels right to put down on the page. The experience wasn't abrupt or abrasive. It began as a yearning within me to go deeper. To examine things more closely—to strain out the noise and opinions I'd picked up over a decade of following Jesus and get back to the root of things.

I'd met dozens of different versions of Jesus over the years. There was the Jesus I met at the megachurches. The Jesus I heard about from that religious group. The Jesus of Instagram. The Jesus who showed up in the Catholic church, who was a starkly different Jesus than the one who appeared in the Episcopalian church nestled next to the brewery downtown. There was Pandemic Jesus. There was Election Jesus. There was N95-Masked Jesus, and there was Anti-Vaxxer Jesus. Jesus seemed a bit like Barbie—so many versions to choose from; scan the shelves until you find the one you want to take home.

And yet, if you had pressed me before the sifting started to occur, it would have been clear I was completely unaware of the nitty-gritty details of what Jesus did when he walked this earth. I knew the grand sweeping motions and the timeline of events, but I didn't know the backstories, the cultural clues, the deep

Jewish roots that shaped him. I knew the rhythms and routines of certain influencers online more than I knew the waking, breathing life of the one I call Savior.

Perhaps this sifting sounds like a faith crisis, but I promise you it wasn't. It was more like deciding to get to know a parent after many years of never asking them questions or bothering to know about the life they lived before you came along. It was the opposite of a crisis in so many ways—it was a stabilizing.

I dared to ask: *Might we meet again, all these years later, and find a deeper level of intimacy?*

This question blazed a path forward for me as I moved in closer to the Rabbi who once lived in a body on earth and walked, healed, and broke bread among his people.

Jesus was a first-century rabbi who grew up in Galilee, a profoundly spiritual place. Galilean Jews held a deep reverence for the faith and memorized the Oral Torah, which was part of a collection of rabbinic teachings called the Talmud. Young men underwent a rigorous education process, and very few would get to the point where a rabbi would be their teacher. If selected, they would leave home and follow the rabbi everywhere, learning how he moved and spoke. They would be his little shadows.

When a rabbi selected a student to follow him, it's believed there was a common blessing friends and family would extend to the student: "*Hevei mitabek b'afar raglehem.*" In English, "May you be covered with the dust of your rabbi."[2]

That was the hope—that the students would follow the rabbi so closely, so intently, that they would be covered in the dust kicked up by the rabbi's sandals as they walked along the roadside together. A layer of dust coating their being. That's how intimate we're talking. When Jesus uttered the words, "Come, follow me," it wasn't a half-hearted invitation. It was always meant to be an all-day, every-day, whole-life kind of following.

In my own study, I scaled back on sermons and other people's insights and focused on the Scriptures alone. I pitched my tent in the Gospels—the four accounts of Jesus's life. I started reading at a slower pace, learning to lean in and notice the rhythms and routines of Jesus.

How he treated people.

His relationships.

The way he used his time.

His compassion and concern for strangers.

His attitude when things didn't go smoothly.

The stories he emphasized and themes he wove throughout his teaching.

What he did with his spare time.

The small things he did on repeat.

There was so much to behold, and new observations came to the forefront all the time.

I'd read a few lines from one of the Gospels and ask, "What can I notice today?" It was my way of following closely—of daring to go beyond always looking for personal applications so I could instead be more curious about the text itself.

I noticed Jesus's lack of hurry. I marveled at the way he took time, even in the most chaotic moments, to make individual people feel seen and known. I leaned in even closer.

Perhaps most striking was noticing the number of times Jesus exited the noise of the world and pressed into secret places. When it wasn't convenient. When the timing didn't seem right. One might say he had his own practice of unplugged hours—using precious time to go away in solitude to pray.

He often retreated at times when I might have said, "Now?

Right now? Just when things are getting good? Just when the crowds want to press in? *Now* you want to get away from it all? This is your moment!" The parts of me that still ache to be seen don't understand his decision-making process. But still, he made the space. Even when things didn't go according to plan, he fought to enter the stillness.

It's a hard prescription for those of us who have long correlated faith with doing. When I found faith a decade ago and put down roots within a church, I fell in love with all the moving and the shaking—all the ways the Bible felt like a Nike ad. I loved the parts about producing things. About running your race. About winning the prize and cutting off all hindrances. I loved knowing faith is something you do.

I was utterly entranced by the words of Isaiah: "Here I am, God. Send me."

I wanted to be God's little soldier, not his child.

For so long, my prayer was, "Use me, God."

But then I started to realize that "use me" was the wrong prayer entirely. If I think back to the people who've used me, those people had motives. They had an agenda. They didn't value me—they wanted what I could offer them. And when I strip away the veneer, that's not love; that's a transaction.

I didn't realize I was slowly losing intimacy with God because I was caught up in all the momentum, resisting the stillness—and the isolation I thought would come from that stillness—at all costs.

———

If Jesus needed to create space in his days to retreat and press into stillness, how much more do we need to do the same? And what might be waiting for us there?

I know that probably feels like the hardest part in all of this. The "getting alone" part. We don't know what we'll uncover if we enter into that vulnerable space.

If you feel this, know you're not alone. Seventeenth-century philosopher Blaise Pascal once wrote, "All of humanity's problems stem from man's inability to sit quietly in a room alone."[3] He wrote these words in 1654. Evidently, sitting still has always been hard for us. We've always been looking for distractions that usher us away from stillness. And yet, God said, "Be still, and know that I am God" (Psalm 46:10). Notice the sequence. The stillness comes first, and then through the stillness comes the knowing. *Be still and know. Be still and know.*

I resolved in my spirit that I would not miss out on the sweet chance to know God just because the "being still" part felt too difficult. So I started sitting still. It's a rhythm I've begun building into daily life—and not just during quiet time. I practice stillness before bigger tasks, before every writing session, and when I find space in the evening hours. Each moment of stillness offers me a chance to step in sync with God—a little touch point amid the busyness of daily life.

I turn off my phone or tuck it away. I situate myself on the couch in my office. I pull a blanket over my legs. Sometimes I light a candle. And then I wait. I'm not waiting on anything specific, but I'm learning there's power in letting my mind meet my body—in waiting for both to still. I let the chaos of my brain trail off—sometimes it tapers off quickly and sometimes I have to wait longer for the mental chatter to cease. I might be tempted to think I have things to do that are more important than learning to sit in the stillness, but I wait here anyway—undistracted and available.

In these spaces of stillness, I often find myself writing down one of a series of questions:

What do you want me to do today?
Who do you need me to see?
What do you want me to know?

I have my own ideas—my own extensive mental lists—but the thoughts that emerge through learning to listen in the silence are always better, and I write them down on the page as they make themselves known.

Write a card to Lane today.
Practice patience with Novalee.
Go about your work unrushed and unplugged.

The ideas that emerge are tasks that could easily go undone, swapped out for something more pressing. They're typically tasks of presence. But each one on the page is a reminder that my time is already going to go in all sorts of directions today—I might as well direct it toward the right things.

Reset the space around you.
Come back to this quiet place tonight.
Send a text to encourage Melissa.

These tiny things won't shake the planet, but they feel close to the heart of Jesus—to the rhythms of his life I'm learning to follow closely—and they feel like life-giving ways to add a little more light into the corners of the world I can touch.

Before I began this practice of stillness, everything felt urgent. I felt I had to use every spare moment to fling myself into the next task. Time was never on my side, and any chance to meet with God had to be filled with divine and booming purpose—with some sort of clear application for myself or someone else.

But as time goes on and stillness fills the cracks in my day more and more, I'm releasing the grip I once had on the transactional aspects of my faith. A gentle intimacy is emerging in their place. The best way I can describe it to you is this: It feels like driving home in the dark of night, pulling down your street and into the driveway, and seeing that someone left the light on for you. You're home. You're safe and someone wants you to be here.

I no longer feel the need to make my faith look flashy or impressive. I'm not concerned with being right or having the best nuggets of wisdom to share. My faith has returned to being what it was when I first reached for it: a lifeline.

And so it feels appropriate to end these thoughts with a blessing for you, my fellow traveler. I hope you take it to heart, because when you do, it may revive parts of you that you didn't even know needed a revival.

Here it is—take it as you go: *May you be covered with the dust of your Rabbi.*

May you be that close.

chapter twenty-two

SABBATH

_____

I often find myself circling back to one of the first invitations Jesus ever extended: "Come to me, all you who are weary and burdened, and I will give you rest" (Matthew 11:28).

Weary and burdened? Yes, and completely. Have you seen the news lately, Jesus? Have you driven on the road with other humans recently? Have you sensed the shift in the air lately? It feels like everyone is tense and on edge. *Weary* would be an understatement at this point.

And yet, the thing about Jesus is that he always asked questions to which he already knew the answers. That was sort of his shtick on earth. He knew about the deep soul weariness that draws us down. He knew we would try with everything in us never to pause, break, or slow.

This invitation to rest tells us a lot about the heart of Jesus. He could have extended any kind of invitation. An invitation to perform. An invitation to measure up. An invitation to puff ourselves up and try to prove ourselves. Instead, he went straight for the bullseye, straight to our worn-down souls, and invited us to step away from the constant racing and uncover deep, soul-restoring rest through him.

And here's the thing: It's just an invitation. It's not a command or a pressure-filled suggestion. He didn't demand. He didn't force. He simply extended the invite, knowing we decline invitations all the time. He extended the invite, knowing that to receive the rest most of us are craving, we'd have to wave our little white flags and just admit it: We're tired. We're worn down. We're ready for a different path.

———

Before the pandemic, practicing the Sabbath was a centerpiece in my life.

If you're unfamiliar with the rhythm, the Sabbath is a day of rest traditionally observed at the end of the week. It is one whole day—twenty-four hours within a 168-hour week—reserved for restoration, delight, and the practice of ceasing.

In his book *The Sabbath*, Rabbi Abraham Joshua Heschel writes, "Six days a week we wrestle with the world, wringing profit from the earth; on the Sabbath we especially care for the seed of eternity planted in the soul. The world has our hands, but our soul belongs to Someone Else."[1] I think that's the worthiest reminder we could have for a twenty-four-hour reset—the world has our hands, but our soul belongs to Someone Else.

A core part of my Sabbath practice was ceasing my engagement with everything digital. I would power off my devices or place them on silent throughout the weekend. I would delete social media apps and take a break from the headlines. This was one of my first real tastes of what the unplugged hours could do for an overdriven soul.

But when the world shut down, my Sabbath rhythm was the first thing to come undone. I knew I needed it more than ever, but I was turning to my phone for answers and hope, so how could I possibly take a break? I convinced myself that if I unplugged, something crucial would happen and I would miss it. I remember many wee morning hours when I lay in bed reading articles on the state of the world, driving myself deeper into morbid rabbit holes of information that never offered me peace.

I know I wasn't alone in that. During the shutdowns, *Quartz* reporter Karen Ho developed a regular rhythm of posting on Twitter between 11:00 p.m. and 1:00 a.m., reminding people to quit the doomscrolling and go to bed.[2] Having a device at our fingertips that could offer us so much intel—a little digital powerhouse—made it easy to believe that if we read enough, consumed enough, and knew enough, we'd somehow get to the other side quicker.

I kept thinking maybe the next bit of information would be the thing I was searching for—a way to loosen the tightly wound ball of panic in my chest.

———

When I started the unplugged hours on my thirty-third birthday, returning to my regular Sabbath practice came with it.

The first week I attempted to bring back the rhythm, I wrangled the whole family into it. I went all out in dramatic flair. I cleaned the house for the Sabbath. I found a Sabbath prayer online and printed it out. I made Lane and Novalee join me in a circle as I lit a tobacco-and-patchouli-scented candle from HomeGoods and recited the prayer out loud. I wanted the moment to be symbolic and weighty. I wanted them to understand that their lives would be transformed at that moment.

After the prayer, we all stood there for a few seconds, wondering what to do next. We never did that Sabbath ceremony again.

That's what I had to reckon with in those first few weeks of reengaging Sabbath: releasing my grip on the perfectionism I was tempted to slather all over the rhythm. I practiced not checking emails, not listening to the hiss of guilt that said I should be doing something more productive with my time. I practiced ceasing. Ceasing work. Ceasing tech. Ceasing productivity. Ceasing striving. Ceasing the need to control everything.

And slowly, I came to the realization that Sabbath is not about control. It's the opposite. It's about learning to release all that control you've been white-knuckling for so long. It's about learning to let go of the narrative you've told yourself so many times—the one in which it's always on you to hold it all together.

———

These days, I'm a student of Sabbath, and I'm learning new things all the time. About myself. About the nature of a God who built rest into his blueprint. About how sometimes a good gift is right in front of us, but we don't know how to choose it because to grasp it, we'd have to let go of something else.

Sabbath is my way of telling myself, *I am not above receiving a good gift.* So I step into it at the end of each week. I can hear the siren calls of busyness, but I choose to resist them and power down instead. I delete the apps. I defy the near-constant itch to check emails or tackle a work project. *It will be there when you get back*, I remind myself repeatedly. *Rest. Don't resist the rest.*

In these precious Sabbath hours, I find myself being infused with the strength to show up for the week ahead. I discover fresh inspiration and new ideas blooming. Over the course of the twenty-four Sabbath hours, my posture shifts toward one of peace and readiness. I went without feeling that peace for years, so I know there's something to this.

I read books and feed my soul throughout Sabbath. I linger in my time with God. I take long walks and long baths. Sometimes I bake. Sometimes I take naps. Lane gets up early and grabs coffee from the local coffee shop. We lounge in bed. We let the countertops fill up. We visit the library and stock up on books. We do our quiet time as a family in the living room, and that usually turns into bacon sizzling on the stove and pancakes on the Blackstone. Sabbath is slow, and while it's challenging with the demands of a toddler, we are trying to model for her the art of resting and delighting in that rest. I think she's getting it in bits and pieces.

As a family, we had to ask ourselves, "What are the sparks that restore us and bring us back to life after a long week?" We practice those things—trips to the local farmers market, a slower pace, spontaneous decisions, extra coffee, concerts, park dates, and presence. Presence, along with the cultivation of a deep appreciation for this life we're living, has become the deep, thudding heartbeat of our Sabbath.

We gather with friends. We cook or order takeout. We play games. We curl up on the couch and watch a movie together. We watch our cups miraculously fill back up.

There are plenty of times when I don't want Sabbath to end—when I don't want to return to the buzzing of life. That's something to be aware of when you start practicing Sabbath—or when you return to it, as I did: The world won't stop spinning. People won't slow down. The churning won't cease. If anything, the noise ramps up. The forward movement only quickens when we step out of the race.

We tend to think that if we leave the race for just a moment, we'll lose our footing and be unable to keep up. That's the fear that keeps most of us going full-speed at all hours of the day. But when we leave the race to embrace rest, we step back into the bustle with new strength. Otherworldly strength. We run even better and with a completely different cadence—one rooted in deep rest rather than deep striving.

We become different versions of ourselves when we clear the space to rest and honor the truth that empty cups need to be refilled by sources bigger than themselves. Only then can they serve their true purpose without wearing out.

My only regret is that I missed out on this refreshment for so long. I went years without resting because my identity hinged on being the person who never needed to take a break. It was a source of pride—but one that kept me running on fumes. I believed that the busier I was, the less time I had to take a break. The opposite is true, though. The busier we are, the more we require that sacred rest to restore our souls. To come back to the heart of things. To remember why we started. To reacquaint ourselves with a power that is bigger than us. To release that ever-tightening grip of control once again.

If I could travel back to that younger version of me—the young woman who believed everything would fall apart if she ceased her striving, I'd tell her what I know now: "You're pouring from an empty cup, love. Don't you want to refill? You don't have to be afraid to rest. Rest isn't weakness—it's the secret to so much strength you've yet to tap into. You can take the break. You can cease. You are allowed to take the weight of the world off your shoulders. No one is asking you to hold it. So go ahead—let that weight roll off you for good."

chapter twenty-three

WATCH

In the days when we were first struggling with Novalee's unexplainable illness, we had no way of knowing it would last for ten months. The first two months were marked by doctor visits, Sea-Bands and saltine crackers, hundreds of Google searches, baby vomit, yanking out the car seat to hose it down yet again, and a lot of untamed anxiety. Just to try to go anywhere, we had to wrap her in a smock and cover her in towels. The whole setup was pitiful.

When we returned home one day after getting more supplies (and mopping up yet another round of baby vomit in the car), I wearily set the shopping bags on the kitchen floor. I felt so exasperated with not having any answers yet. We were stuck in the middle of what felt like an ongoing storm.

I watched my newly walking daughter totter over to the shopping bags and pull out a box of saltine crackers we'd opened on the car ride home. She plopped down on the floor, pulled out an already-opened sleeve of crackers, and poured them all onto the floor before shoveling them into her mouth. Since she was hardly able to keep down any food at the time, this was her feast.

I sat down on the hardwood floor a few feet away and watched her, knowing she was unaware that so many people were worried about her. She stood up, grabbed a fistful of saltines off the ground, and clip-clopped her way over to me. Stray crackers crunched beneath her sneakers. She plopped down in my lap and continued eating. I picked up some crackers off the floor and joined her.

We'll clean up the mess later, I told myself. *Who cares? Right now, I will stay in the middle of this mess and figure out how to live in it.*

———

A few mornings later, I found myself at the dining room table with a journal open and my Bible beside me. I was doggedly flipping through pages after feeling a single word drop into my spirit like a divine download: *Watch.*

I think this happens in our faith sometimes—we sense a word making itself known within us, and then that word starts popping up everywhere. Sometimes our understanding of the word's significance is instant, and other times it unfolds gradually and in layers.

Watch feels like a pretty appropriate word for someone in the midst of an unplugged journey, but the term also felt cloaked in mystery to me. It was a word I'd never really noticed in the Bible before, so I started tracking every instance in which it appears across the Scriptures, attempting to decipher it like a code I was meant to crack.

I discovered that *watch* is often paired with prayer throughout the Bible. *Watch and pray. Watch and pray.* I wish I could say prayer is my first reaction when hard circumstances arise. When I'm caught up in venting all my problems and dramas, I so admire the people who stop everything and say, "Let's pray." *Why didn't I think of that?* Reactions I typically have before prayer hits the docket include wallowing, crying, freaking out, venting, worrying, panicking. But the words *watch and pray* strung tightly together feel like a whispered melody for an anxious mind. When in doubt, watch and pray. When in the storm, watch and pray. When in mourning, watch and pray.

A request to "watch" came from the lips of Jesus soon before his crucifixion—right before the guards arrested him and the end began. In Gethsemane, he asked his disciples to "keep watch" while he went into the garden to pray, but they couldn't do it. They kept nodding off and falling into slumber (Mark 14:32–42).

The Greek word translated "keep watch" in Jesus's Gethsemane statement is *gregoreo*. It means "to be vigilant; to keep awake."[1] When this term is used in the Scriptures, it's often a nod back to the soldier's night watch, where watchmen would stand at their post and be on high alert for any signs of danger on the horizon. They would stay on guard, determined not to miss a thing.

Another definition of what it means to watch is "to keep vigil." That's what Jesus asked of his disciples: Stay here and keep vigil with me. This idea entrances me. When we think of vigils, we often conjure up images of candlelit prayers and services of remembrance. But keeping vigil in this instance is the deliberate practice of staying awake at a time when it would be more common to sleep. It is a "period of wakefulness."[2]

Heather Hughes, a publication specialist from the Center for Christian Ethics at Baylor University, writes, "Keeping vigil engages our natural bodily response to moments of intense love, fear, sorrow, compunction, or awe. . . . Our sense of what is truly important impels us to be fully present, without seeking distraction or escape."[3]

While it's easy to imagine keeping vigil as a physical practice, I believe it's also possible to keep vigil spiritually. When it would be easier to sleepwalk through our lives or distract ourselves from experiencing the pain that comes with being human, there is immense power in doing the opposite—in choosing to stay awake and alert to what is happening within us in storms and in sorrows.

There have been so many times when I've walked through hard circumstances in a state of dazed distraction, but in this season of Novalee's illness I felt God standing by me in the mess and saying, *Something is happening beneath the surface. You don't see it yet, but keep watch. I'm weaving a precious golden thread in and out of this story. Stay awake for all of it.*

I didn't know it yet as I sat there with my Bible and my journal, but later that day we would rush Novalee to the emergency room. *Watch* would suddenly take on new meaning as we waited on doctors and nurses and answers. Through the testing, through checking yet another possibility off the long list of potential diagnoses, through the visits with specialists from all over the hospital, I felt strangely steady and still as I whispered to myself, "Watch. Watch. Watch."

———

One afternoon, wedged between two hospital stays, I was on a call with my therapist, and I asked her about "processing" what was happening to us. I was walking through our neighborhood as we talked. There was finally an autumn chill in the air, and the leaves were starting to fall. The season was turning over.

"So many people keep asking me how I'm processing everything, and I don't know what to tell them," I said. "What does it mean to process? How do I do that well?" I wanted a concrete response to my question—some way to measure my progress. But my therapist's response felt more like liquid than concrete.

"You remain present," she said. She went on to explain that some people distract themselves away from the present moment when it's hard to face—so the details are scattered and confusing when they go back later to process the events. "Being present to what is happening is the key to processing," she said. "You remain present."

You remain present. You just stay in it. Even though you don't know where you're going or when your situation will end, you keep walking forward and remain present for all parts of the story.

Her advice reminded me of a book I'd read a few months earlier by Henri Nouwen. In it, he suggested that maybe the idea of "picking up our cross" was a call to stand inside the hard thing we're facing and not run from it or distract ourselves. He wrote that maybe picking up our cross is inviting God into the hard thing and realizing it won't destroy us.[4]

So that's how we operated throughout that season. We kept rooting our heels in the mess and inviting God in. We stayed on watch, but maybe more importantly, we learned to let others come in and be the night watch for us.

In the moments when our spirits dipped low, others stepped in and helped us move into the next day. People sent meals. People checked in. People dropped things at the door. It was like a relay race was operating in the background of our lives—the metal baton being passed repeatedly. People kept being tapped into the race, pumping us full of courage to keep going. People rallied. People prayed. Throughout that whole season of hospital visits and doctors and getting no clear answers, we somehow felt continually covered in a cloak of unspeakable peace. I can only believe this was because of other people's prayers—those who kept vigil for us until we safely reached the other side.

———————

The Quakers have a saying, "I will hold you in the light." It's a way of telling someone, "I am talking to God about you. I am uttering words to him even though everything around you feels so dark and dismal. I'll keep placing you in the light."

I love the image of placing our friends and family in the light when they feel broken and tired or when they don't have the capacity to talk to God themselves. It reminds me of that story in the Bible about the four crazy friends who were adamant about getting their paralyzed friend into the presence of Jesus (Mark 2:1–5).

In the story, Jesus was preaching to a packed house. Imagine people piled on top of one another, making it impossible for anyone to leave or enter the home. And while it's easy to imagine this was a crowd of adoring fans, the reality is that a lot of these people were likely spectators and skeptics. They wanted to get a glimpse of this new rabbi, but there's no evidence in Scripture that they were completely sold on him.

And yet, the four guys who couldn't enter the house because of the crowds were determined to get their one suffering friend to Jesus. They were not skeptics but light-bearers. When they couldn't get into the house in the conventional, through-the-front-door manner, they got scrappy and lowered their friend down to Jesus on a mat.

Only recently has this story started hitting me like a ton of bricks. I'm forced to ask myself, *Am I someone who would claw through the roof of a house for my friend, or am I more of a cautious spectator?*

I'm enamored by the beautiful truth that there are people in this world who are so present and awake that they will claw through a literal roof for the people they love. I think that's what each hour of presence through the unplugged hours is revealing to me: a deep desire to hold my people in the light—through prayer, through love, through physical and spiritual presence, through vigil.

We could be the light-bearers. Imagine that. We could be the ones to bend low and say, "I'm talking to God about you. I'm holding your tired body and soul up into the light."

We don't have to know all the answers to sit close to someone in their agony. We don't have to have perfect sentences to start mumbling words to God. We don't need anything we don't already have to practice what I'm calling "uncommon presence" in someone else's storm—just the willingness to stay awake in those uncommon hours.

In a time when the darkness feels wild and sprawling all around us, we step into the night boldly. And there our watch begins.

chapter twenty-four

I WILL BE WITH YOU

When we left the hospital the second time, three months into our ten-month season of Novalee's illness, the doctor prescribed a daily anti-nausea medication. It wasn't a definitive answer, but it would help the medical team continue to check things off a more extensive list of potential diagnoses.

"I want you to have some normal days. You need that," the doctor said. And what I heard her saying was, "I want to give you the space to breathe again." You know when someone tells you something you don't want to hear, but somewhere deep within, you know you need to listen to it? That was this.

In the months that followed, we still didn't have answers, but we did have a span of beautifully boring days. Days when we didn't have to be on high alert. Days when we could wake up slowly. Days when we weren't cleaning up vomit from the car seat. Days when there was no reason to Google for the one thousandth time. And we all released a collective exhale.

Six months later, when it was time to start weaning Novalee

off the anti-nausea medicine, my anxiety spiked once more. We had an incident when she got sick on the way to breakfast—the first of its kind in months. I texted the girlfriends in my daily Bible reading text thread and asked them to pray. My mind was frantically returning to old, familiar spaces of fear. Rule number one when that happens: Invite others in.

That's the funny thing I've discovered about technology— the balance is everything. We can do all the unplugged hours in the world and become the most present people in the process, but there are still times in daily life when it helps to know others are reachable—ready to come alongside us to pray, to encourage, to remind us they're a text or a call away—because of the device we're holding in our hands.

I listened to my body throughout the morning. I napped when my daughter napped. I woke up and began to clean. Cleaning is my reset mode. Or maybe it's a coping mechanism. I know I can't control all the elements of my life, but I can control how clean the toilets are.

I didn't turn on a true crime podcast or an audiobook to take my mind off things. Instead, I stayed powered down and chose to pray. I use the word *pray* loosely. I rambled. I sputtered. I spiraled as I pulled bins of Novalee's toys out from under the wooden play table and started dumping them on the floor. As I hung up mini plastic spatulas and cooking pots in her tiny play kitchen, I remembered a story from the beginning of Exodus.

It was the part where God commissioned Moses to lead the Israelites out of slavery in Egypt. He showed up to Moses in the form of a fiery bush as Moses was tending to his father-in-law's flock of sheep. And Moses had questions, as he should have. If God showed up to me by manifesting himself in a fiery bush, I'd have questions too.

Moses's questions were less about the mission and more about himself, though. He had all kinds of rebuttals and excuses and reasons why God should find someone else for the task at hand.

I find this whole scene peculiar. It's easy to forget that Moses, at one point in his life, might have considered himself wildly qualified for the leadership position at hand. If there had been glossy hardcover high school yearbooks back then, Moses probably would have been labeled Most Confident or Most Likely to Succeed. He belonged to two worlds. He had an Egyptian education combined with solid Hebrew roots. But when his reactive sense of justice caused him to kill a man in broad daylight, he was forced to go into hiding for the next four decades.

Something about forty years tucked away in obscurity must have eroded his inner sense of confidence. He went from seemingly having everything to being in a position where no one noticed him. Likely somewhere between the sheep-tending and the slow rhythms of the monotonous daily grind, he preemptively disqualified himself. Herding sheep, changing diapers, caring for a sick parent, or doing any job that doesn't get recognition can do that to us.

And so, Moses asked God the obvious question: Who am I that you would use me?

It's a question I've asked myself no less than a thousand times—and it's a question I continue to wrestle with to this day. It's not that I want to hear God's divine reasoning; it's that I want to be affirmed. I want God to beam down a crystal-clear statement, preferably handwritten, that says, "You are intelligent and brave and ready for this. You've got this completely in your own strength, but tap me in if you need me."

I want to feel qualified. Who doesn't?

We like it when someone sees us, applauds us, and reminds us that we're doing okay.

So maybe that's how Moses was feeling when he asked God,

"Who am I?" Perhaps he had that anxiety we all feel when stepping up to the plate. The worry that says, "Can you just tell me I'm enough for what's coming up ahead? I'd love to be at least a little bit qualified for what's coming next."

God, because he's subtly hilarious and never one to back down from a teachable moment, doesn't answer Moses's question—at least not with the answer Moses wanted to hear. Instead, he answers Moses with five words: "I will be with you."

Not "You have what it takes."

Not "You're completely qualified."

Not "You're the right person for the job."

Just "I will be with you."

He reassures Moses, "You're looking for something that qualifies you to step into what's coming next, and it's right here. It's me. I'm going with you. Of all the things I could offer you, this is the most important thing, the thing that will allow you to be okay in all of this: my presence."

Ah, there it is. Presence. In a year where my goal had steadily morphed from counting unplugged hours to cultivating a lifestyle of presence, I had to reexamine what I believed to be true about God's presence. Did I think it was full and constant? Or did I believe it was stingy and running out?

God's version of presence is not like our presence. It's not, "Let me do this one thing first, and then I'll be with you." It's not the kind of presence where he occasionally glances up from the phone to make sure we're okay. It's highly concentrated and it's complete. It's all-encompassing. It's an "I can't take my eyes off you" kind of presence. Maybe we don't immediately see the power in that—the novelty in that kind of constant contact—because we rarely experience that kind of presence from one another. But when we do feel that sort of presence—when someone looks at us like we're the only one in the room—it makes us feel like we can take on anything. That kind of presence makes us bold. That kind of presence gives us the courage to walk through doors we didn't think were open to us.

I will be with you.

Those words floated into my brain as I sorted through a bin of brightly colored wooden fruits and vegetables. My hunched shoulders relaxed almost immediately when I repeated those five words under my breath—"I will be with you." Something that had been wound tight within me released a little.

"It's not on you anymore." I think that's what God was saying between the lines.

It's not on me, and it's not on you. We can release. We can exhale once again. We don't have to act like we have all the scout badges that qualify us for the season we find ourselves in—or the one we'll walk through next. We have a presence who goes before us and follows behind us. We have a companion who never leaves us.

Instead of the question, "What if God's not here?," I've started to ask some better ones in its place: *What if he's in absolutely everything? How does that change the way I approach each moment? The way I show up? The way I move forward?*

I know I can't silence your fears about the future. I get it because I have those fears too, and they can get loud and cocky sometimes. But we don't have to muster up some veneer of confidence. Maybe we can look at all the roles that feel too big for us—at home, at work, and in all the spaces in between—and repeat back to ourselves, "He's with me. He's with me. He's with me," as we step fully in. Just those three words until we believe them with our whole being. Until we walk taller. Until our fears loosen. Until all the spaces that require our presence are flooded with the divine.

I walked through the process of weaning Novalee off the anti-nausea medication unsure of who I'd be if she got sick again. And while I could have filled that time with worry and fear and doubt, I leaned on a less pressure-filled truth: Maybe I could just be her mom. The one who cleans up the throw-up. The one she falls onto after a hard day—the safe landing place in a world that's brutally hard. I might have to lay down my expectations for how I think things should look, but I know I'm qualified to be her mom. And it's not because I've read enough parenting books or mastered the art of presence on my own: it's because God's presence is with me and in me—running rich and deep in my veins.

That's what qualifies us. That's the promise. And here's the beautiful thing about God's promises: They never break. They're durable. They last.

He's with us—even when we don't feel it. That doesn't mean we get a guarantee of better circumstances. It doesn't mean life doesn't still break our hearts from time to time or that an easy path will emerge in the woods we're walking through. But there's an internal shift. A realization: He walks with us. He covers us. He goes before us and follows behind us. That's the whole pep talk. The pressure is off.

chapter twenty-five

THE POWER OF PRESENCE

The term *Imago Dei* keeps coming to my mind lately. It's Latin for "image of God." "*Imago Dei* means in likeness, or similarity, to God," write the editors of Christianity.com. "Humans are created with unique abilities, absent in all other creatures of the earth, that mirror the divine nature of God."[1]

I imagine this idea is like when you look at a growing child and see parts of their parents emerging—his father's mannerisms, her mother's striking dimples. These traits are markings, clues to who they've come from.

If we are made in the likeness of God, then we have his markings. I picture us as little mirrors who, in our best moments, reflect his goodness and his kindness and his light. We give others a glimpse of him.

When God spoke those five words to Moses, "I will be with you," he promised Moses his presence—the most powerful expression of himself that we as human beings can experience in our daily lives. But if we carry God's likeness, might that mean our presence—our physical presence—holds some power too?

That's a question I've been asking myself in the midst of unplugging; I'm challenging myself to look up and look around instead of looking down. I think about it when I talk with the barista or ask the lady at the checkout how her day is going.

My presence may hold power. There's some weight to the idea. And yet the way our world operates often undermines the potential benefits that could come from it. Instead of presence, the world pushes remote interaction. More Zoom meetings. More texts. More multitasking. More screens.

How do we navigate the tension between believing that our physical presence has power and feeling pressured to digitize our lives to keep up?

———

When I think back over the times I've felt the most seen and understood, the most known and capable, those occasions rarely involved a device. Most of these times involved a table and someone sitting across from me. There has always been another person speaking light, love, and belief into me when I didn't believe those things for myself or needed the reminder. Most of these moments have happened when I've been physically with the person, when I could see their eyes locking with mine.

The first time we went to the emergency room with Novalee, there was a moment when something snapped and broke within me. Because COVID restrictions were in place, Lane stayed in the parking lot while I stood in a long line with Novalee, waiting to be checked in. When we reached the front of the line, a stern guard handed me a form on a clipboard and motioned for me to move to another line. Once we checked in, I found a seat in the corner of the room and waited.

Kids moved and played all around the room. I sat in a plastic chair, and my daughter lay reclined in her stroller in front of me. She couldn't open her eyes or muster up the energy to do anything but lie there. I knew something was wrong, terribly wrong, but I didn't know what to do. We found out later that her sugars were so low she easily could have slipped into a coma.

All the fear, tension, and worry I'd been holding in since she first started getting sick finally bubbled over and I started to cry. Not gentle tears streaming down my cheeks but a heavy, sobbing, "God, help me" cry.

The stern guard who'd given me the clipboard an hour earlier moved toward me. She looked at my daughter, kicked into gear, and brought the nurses in. As things started moving around me, I locked eyes with the guard and saw gentleness and compassion there.

A woman sitting two chairs down swayed back and forth with her hand stretched out over my daughter and me, saying repeatedly, "Jesus. Jesus. Jesus." A prayer.

And though I had no previous connection with either of these women, they were suddenly present with me in one of my darkest moments. I felt cloaked in peace. There was power in their presence—in the God who was no doubt moving within them—and it filled me with new strength. Even though we were all wearing masks that day, I never see the masks in my memory of that moment. I remember only how these women looked me deep in the eyes and chose not to look away.

I know our presence holds power because the presence of others has encircled me at various junctures throughout my life, and it changed everything. It convinced me to stay. It convinced me to fight. It convinced me to keep seeking God in barren places.

———

When I say our presence holds power, I don't mean it in a cute way or even a thoughtful "maybe we should stamp this on a coffee mug" way. I think the power of our presence is far weightier and more important than we realize. It might even have the power to save lives.

Every day, there are people walking around this world who think things might be better or easier without them. And then there are people who decided to stay in this world because of someone else's presence—because love picked them up off the ground and taught them to stand again at a time when they felt completely forgotten.

I heard a story recently about a man who felt lost and forgotten. Deep in depression, he'd gotten to the point where he'd decided to take his own life. And it was a neighbor he barely knew who, in prayer, felt the nudge to write him a letter and drop it at his doorstep. That letter is why he's still here today. I've heard these kinds of stories over and over again across the years, and they always confirm this truth deep in my spirit: The smallest act of noticing someone else, whether they're in the shallows or the waves, can have a profound impact beyond what we might ever understand or imagine.

I want to be careful not to oversimplify, but I also believe that sometimes even the smallest bit of eye contact can bring someone back to life. Sometimes one look delivers a much-needed dose of encouragement, poured like water onto the driest of soil. Sometimes the simplest wave, nod, or acknowledgment can buoy the spirit of someone drowning in a sea of people looking down at their phones, too busy with what's happening in their own worlds to notice anyone else. My neighbor Bebe ends every interaction with the words, "In case no one has told you today, I love you."

I've started saying that too, because the words are true but so rarely said aloud.

People in our lives need a reminder that they're loved on a random Tuesday or a typical Saturday.

People in our lives need to know that someone is talking to God about them.

People in our lives need us to show up at the door, the table, the event, or the party, even though it would be easier to stay home.

We have people who need us to see them today.

With devices always in our hands, we stand to lose the power of presence—our power to see one another, to reach out to someone, look them in the eye, and say, "I think you should stay and fight. We need you here." And though I know I'm not with you right now—our knees aren't touching, and you can't see me from across the table—I'd kick myself a million times over if I missed the chance to tell you the same thing.

I think you should stay and fight. We need you here. I don't know everything you're walking through right now, but I imagine you have days just as I have—days that feel heavy, days when you're tempted to say, "What's the point?" I bet there are days you feel forgotten or isolated or alone. I have them too. But I also know, without a doubt, that there are people who need you here.

Good things are waiting for you up ahead—even if you can't envision them right now. There are sunsets you've yet to see. Good coffee to drink. Art to make. Words to be written down in journals and margins and cards.

There are baby cheeks to kiss. There is food to enjoy and laughter yet to erupt. There are conversations to be had. Blessings to be given. Miracles waiting to take place.

There are people you haven't met yet who will come into your life and shake up your whole existence with their presence. Just wait. It's going to blow your mind.

There are songs to sing, hymns to hum. There are important tears you'll need to cry—goodbyes you'll need to say.

There are arms you will lie in, tables you will sit around, and hills you will climb to witness the view on the other side.

There are prayers to be prayed. Paths to be crossed. Books to be read. Stories to be heard.

There is more. There is more. There is more.

chapter twenty-six

THE INNER LIFE

─────────

We live next to a house that has been unoccupied for over thirty years. The windows and doors are barred and boarded up. The roof looks like it might cave in at any second. The backyard is a forest of sorts—a thick tangle of weeds, poison sumac, and overgrown ivy. For as long as we've lived here, no one has set foot inside that house. Day after day, it just sits there. And it waits. And it continues to fall apart.

Every few months, the owner sends a landscaper to trim the hedges and mow the grass in the front yard. The other day, I watched the owner himself pull up in front of the house with a truck full of supplies. For the next two days, he stood outside the dilapidated house holding a bucket of white paint and slowly, steadily repainted the entire exterior.

I watched as he rhythmically moved the brush up and down, seeming to be in no hurry. I wondered if he thought the paint was actually doing anything for the house. It looked a little better when he was finished, but you can't walk by that house without wondering what's going on inside. Maybe that's why he showed up with the white paint in the first place—where do you even begin when something needs so much work?

I think a lot of us are more like that homeowner than we realize. It's easier to work on what's outward-facing—the looks, the job, the success—and stick relatively close to the surface when pursuing any kind of change or transformation. Everyone loves a good makeover—especially if it can play out in sixty minutes or less. Why wouldn't we opt for improving the external when the internal is harder to work on and rarely ever seen?

We all have an inner life, whether we realize it or not, and there are many ways to describe it. Pastor Gordon MacDonald called it our private world. Christian philosopher Dallas Willard wrote about tending and caring for our souls. Summarizing Willard's perspective, author John Ortberg wrote, "If your soul is healthy, no external circumstance can destroy your life. If your soul is unhealthy, no external circumstance can redeem your life."[1] I sometimes call this soul space my inner room because I like the idea of a physical space within me that feels vital and grounded, a place I can return to again and again.

Catherine Doherty, a Russian-born Canadian spiritual teacher and activist, often referred to the inner life as *poustinia*, the Russian word for desert. We tend to imagine deserts as barren places where nothing can grow, but deserts are often fertile ground—places with immense possibilities for growth once water pours in.

Whatever term is used, the inner life is secret and tucked away. It encompasses all the places within us that we don't give the world access to. In a connected age where popularity, success, and relevance are often our currency, it's easy to neglect the inner life. If no one can see it, why bother investing in it? But when we focus on the outer life alone, we inevitably fall into what leadership coach Susanne Biro refers to as the "outer life trap," where we begin to define ourselves by external circumstances and look to people and life events to keep our sense of validation intact.[2]

Instead of asking, "Why bother with the inner life?," Gordon MacDonald poses a question with expansive potential to impact the shape of our lives: "Are we going to order our inner worlds, our hearts, so that they will radiate influence into the outer world? Or will we neglect our private worlds and, thus, permit the outer influences to shape us? This is a choice we must make every day of our lives."[3]

It's possible that by doing the work necessary to turn inward, we'll uncover treasures within us we never could have anticipated—a profound sense of empathy, a burgeoning creativity, an untapped brilliance, a path toward healing, a kinder voice that tells us, "New things are coming. It's time to rise up."

———

I was just two hundred hours shy of meeting my goal of one thousand unplugged hours when I discovered how barren my inner life had become—and it happened on a dude ranch in Colorado, of all places.

I was surrounded by other inspiring women from my industry and probably could have made some deep and lasting connections, but I spent much of the three-day trip stuck in my own head, battling deep insecurities. To be clear, no one was unkind to me and there was nothing wrong with the trip itself. The problem was internal. I felt like a shell of myself on that trip, but it was on that dude ranch that I learned this essential truth: We can be completely unplugged and yet still struggle to be present because of the voices in our heads. The voices that tell us we don't belong.

The voices that tell us other people don't want us around. The voices that tempt us to play small.

I flew home from that experience with new resolve: *No more. I will not keep letting these insecurities win, stealing precious life from me. I have to stop waiting for other people to make me feel valuable. I will not spend the rest of my days battling these lies.*

If there was an instruction manual for cultivating an inner life, I imagine the first step would be something like this: Be willing to embrace the profound and sacred work of getting alone with yourself and figuring out what needs to change.

You might discover you have anger welling up inside you. It's been left unchecked for years, but it keeps spilling out onto your husband and kids. You might realize that you keep making promises to yourself and breaking them, and that pattern—the one of never showing up for yourself—is breaking you.

There might be lies to face. Fears to address. Crazy thoughts to rescript. You might have a whole airport conveyor belt's worth of emotional baggage you've hidden away for decades, or you might find your inner world is decrepit or covered in a layer of dust because anytime you have the chance to look inside, you pick up your phone instead.

Once you've examined your inner world—once the ground there is broken up and cleared of the overgrowth and debris that have accumulated over time—you can start to build something new there. Something enduring and resilient. Something vibrant. Something peaceful. Something far sturdier than the fear and doubt that never gave you a solid footing.

Author Anne Lamott quotes a friend who said, "My mind is a bad neighborhood that I try not to go into alone."[4] The doer in me would like to add an addendum: "My mind is a bad neighborhood that I try not to go into alone. But I'd like to volunteer to be on the cleanup committee."

That's me these days—donning the bright orange vest, holding a trash bag, and scouring my mind for the litter I've let pile up over time. The lies I was believing. The anger I wasn't facing. The fear of irrelevance that kept me moving yet also made me feel wildly insecure for too long. And while I can't lend you the unabridged version of what building an inner life has looked like for me, I can provide some CliffsNotes that might help you in your own process.

I started by silencing a lot of the extra noise in my life by drastically reducing the amount of media I was consuming: the reality TV, the Netflix binges, the constant stream of podcasts. From there, I carefully and slowly added a few wise voices back in.

I ushered in more prayer, more gratitude, more stillness, and more celebrations of victory—no matter how small. I started reading books on the inner life and mental toughness to see how I could change the landscape of my thoughts. I began taking literally the apostle Paul's admonition to "be transformed by the renewing of your mind" (Romans 12:2). One by one, I started pulling the chaotic lies out of my mind and pinning them down onto pages in my notebook instead. And one by one, I started crossing them out and drafting better truths to stand in their place. I told myself, *You only have to believe these truths just a little bit. Mustard seeds are welcome here. Stronger belief will come with time.* And I was right—my belief became stronger, right on time. I powered down more and more and turned to a notebook and a quiet space when I felt something deeply. I became curious about certain narratives I'd internalized, and I processed (read: am still processing) a lot of them with my therapist. I started to face my feelings head-on rather than turn to my phone to pacify me or convince me to do nothing at all.

And lest I give the impression that you must get away for hours to tend to your soul in secret places, let me tell you the truth: That's not the case. Even if you have only fringe hours and fringe minutes to dig deeper—small pockets of time between work and family and the demands of your everyday life—good work can happen right there.

Once you stop telling yourself that you don't have time for internal work, the growth will begin happening in seemingly random moments. At the grocery store. While cleaning. In the in-between moments when you're tempted to grab your phone or turn on a podcast. In your willingness to stand still for a moment or two and just be.

No, the work isn't instant. I'm afraid it's ongoing and never-ending. Whenever I think I've completely dealt with something, a new weed sprouts up and I have to tend to that too. But the work is good, and the results compound over time. Looking inward and deciding to face what lies beneath the noise has become a steady rhythm. I like who I'm becoming in the process—stronger, more patient, more grace-filled—so I keep returning for more.

Lately, the word that keeps coming to me is *vibrant*. That's what I'm chasing. I like how author Carol McLeod defines it: "A vibrant person is *alive* in the deepest part of his or her soul."[5] Yes. All of that. More, please.

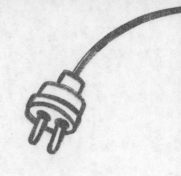

part four

AND BEYOND

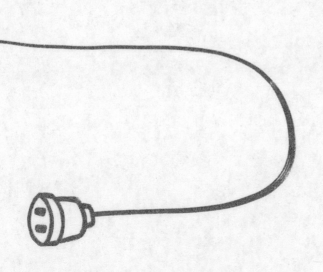

chapter twenty-seven

PACE

A pile of brightly colored envelopes sat on the edge of the kitchen counter, taunting me from afar. They'd been piling up the entire holiday season.

Before the Christmas season even began this year, I felt like I couldn't keep up. This is abnormal for me. I'm typically the one blasting Mariah Carey's high notes at midnight on Halloween, but this year everything about the season felt like it moved too fast. I kept asking my friends, "Does it always feel this way? Why do I feel like I can't keep up anymore? Is this the start of old age?"

Because I was still resistant to slowing down, the closeted control freak within me focused on something she felt she could control—the Christmas cards. I declared with a vengeance that we would not be sending them. We would not be participating in this holiday tradition just because everyone else was doing it.

The matter should have stopped there. No Christmas cards.

Good for you, Hannah. Moving on. But it didn't. Not sending out Christmas cards turned into not opening a single card that arrived in our mailbox all season.

Card after card ended up on the countertop, still in their envelopes.

A pile formed.

Each card stirred unexplainable, Grinch-like rage within me. *How dare these people have their lives so under control that they send these cards to us!*

Lane observed the pile getting higher and higher every day. He dared not open a single envelope, though he knew fun pictures of his friends and their kids were waiting inside. He didn't question my motives.

The pile grew.

I opened the bills. I opened unassuming mail. I opened coupons. But I took every remotely holiday-esque card and threw it in the pile unopened.

The pile sprawled.

If I had discussed the Christmas cards with my therapist, she probably would have helped me see that I was feeling insecure. To me, it seemed like everyone else had no problem moving at a breakneck pace. And because I've already bared my soul to you about my Christmas card identity crisis, I'll also tell you this: The unplugged hours were to blame.

When I started unplugging, I thought the impact would be confined to the surface level of my life. I thought I'd gain more time and be more present. But as I kept tracking the hours, the "move faster" mentality that had always been my inner compass started to break. And I panicked. I didn't know who I would be if I stopped moving at such a relentless pace. Instead of finding out, I said yes in all directions.

Yes to more events.

Yes to more last-minute projects.

Yes to more work.

Yes, yes, yes. Let's keep this train rolling forward.

When December 27 hit, I was mentally frayed but flooded with relief that Christmas was finally over. We cleaned the house in preparation for the new year. I took the Christmas cards off the counter and prepared to drop them in the trash. Lane shot me a glance as if to say, "Really? Are you really about to be so petty? Please call your friends immediately and tell them to stop wasting good stamps and paper on you."

Begrudgingly, I stepped away from the trash can and placed the pile back down on the counter. Slowly, I opened one card. And then another. And then another. I spread out the cards across the dining room table. I stared at them, suddenly feeling silly and petty. I loved all these people. I love their faces and that we send something tangible to one another once a year. Amid all the fast-paced traditions we've adopted, this slow one still holds. We still buy the stamps. We still address the envelopes.

A feeling of joy and gratitude spread over me. *These are our people. How blessed we are. How nice it feels to survey their faces— the smiles, the aging, the growth of their kids—in my physical world rather than on a screen.*

And that's when it hit me—the "I need to keep up" mentality was my own doing (or maybe I should call it an undoing). The pace I had chosen was the reason everything felt so rushed and hurried. I'd taken all the elements of the season, activities that could have been fun and spontaneous—and optional—and turned them into a to-do list of robotic motions to go through. It's no wonder the whole season felt drained of meaning.

Something had to give. I didn't necessarily want to slow the pace, but I was afraid of what might happen if I didn't. I've visited the pit of burnout once before. The view wasn't worth the admission price. I've sat in severe depression from spreading myself too thin. I'm not going back to that place again.

I took our bulletin board down from the wall—the one we reserve for invitations, artwork, and reminders—and cleared it off. I filled the whole thing up with Christmas cards two days after Christmas. It was a tiny act of resistance—a way to remind myself that I'm determined to be done with trying to keep up with the world around me. The cards stayed up there, proudly hanging, until late March.

———

The more I press into the unplugged hours, the more I've been having random childhood memories about my mother. They keep popping into my brain at different times of the day—and I don't fully understand why, but I have hunches.

The details and specific circumstances are often fuzzy, but I know the memories are from different places. Sometimes we're moving through a department store. Sometimes we're walking on the beach. But it's the same scenario on repeat: I'm somewhere with my mom, and she's taking forever to get on to the next thing—at least that's what my little brain was thinking at the time. She'd be moving so incredibly slow—if we were outdoors, she'd be stopping to take pictures of random bits of nature with her Kodak film camera. I already knew that when the film was developed, we'd have to lay the grainy photos on the kitchen table and try to decipher what she meant to take pictures of in the first place—obscure patches of grass, images taken way too close-up, birds, lots of birds.

Her stopping and starting and relentlessly slow pace drove me mad as a kid. But as the memories surge back, I'm seeing things differently. It's as if I needed time, space, and a lot of life to happen before I could understand what my mom was doing all that time. She didn't have her head in the clouds, as I'd thought. She wasn't distracted—she was just being attentive to the world around her. She was taking things slowly enough to appreciate the beauty around her. She was curious. She was lingering. She was setting the pace for her life, refusing to go faster just because others around her were hurrying.

———

The other day, I rediscovered the story of Elijah.

Elijah's story is one I've read many times before, but something about it is hitting me differently in this season. The stories feel like a feast—a table set with all the things I've been craving.

Peel back the layers and Elijah's story is about learning to rest and allow God to restore us. It's about learning to slow down and set a different pace after pushing the gas pedal too hard for too long. It's a story about finding God outside the wall of noise we've built up to surround us like a fortress.

If you can, go and sit with his story. You'll find it tucked in 1 Kings 17–19. It holds so much texture, nuance, and unexpected softness.

Here's what arrests me: Elijah lives a pretty front-and-center life. I would say he has a flair for the dramatic. He's used to the big miracles, the flashy scenes. But after one big miracle comes a downward spiral that most scholars deem to be a pit of depression.

Meeting him in one of his darkest moments, God gently pulls Elijah out of the spiral. He shelters him under a broom tree and nourishes him. Through angels. Through fresh bread. Through jugs of water. Through presence and naps. The scene feels sacred and tender; I felt like I was noticing a side of God I hadn't seen before. The words *tender* and *God* are two I likely wouldn't have put together before this.

A little later, Elijah is shaken awake by an angel who tells him to get up and eat again. There is a long journey ahead. "Nourished by that meal, he walked forty days and nights, all the way to the mountain of God, to Horeb" (1 Kings 19:8 MSG).

I pull up the locations on Google Maps and trace my finger along Elijah's route. I realize it could have been completed much more quickly than forty days. A person could make the journey on foot in approximately eleven to twelve days. So why all the extra time? Why the painfully slow pace?

Some scholars assume the terrain was rocky; others remark that what matters is the number forty. This number indicates it was a profoundly spiritual journey in which God set the pace for his traveler.

I think about all the care God must have embedded in that slower journey. All the stops to nap. To nurture. To keep Elijah pushing forward in the face of his doubts.

I imagine all the temptations Elijah may have faced to move at a quicker pace. To go back to pushing the pedal. I wonder if Elijah ever embraced the slower pace, maybe even found a little pleasure in it. When I'm befuddled by a slow pace that seemingly makes no sense, I keep coming back around to this thought: Don't rush the process.

I think "don't rush" could apply to many parts of our lives: Don't rush the season. Don't rush the learning curve. Don't rush the waiting. Don't rush the healing. Don't rush.

We often want to rush because we've become used to the efficiency of our digital age, but there are many things in life that cannot be rushed if we want them to turn out right. Many aspects of transformation require being willing to simply take things one slow step at a time. I like to imagine that's what happened to

Elijah during his long walk—a transformation that couldn't be rushed.

I love the way theologian Kosuke Koyama put it:

God walks "slowly" because he is love. If he is not love he would have gone much faster. Love has a speed. It is an inner speed. It is a spiritual speed. It is a different kind of speed from the technological speed to which we are accustomed. . . . It goes on in the depth of our life, whether we notice or not, whether we are currently hit by storm or not, at three miles an hour. It is the speed we walk and therefore it is the speed the love of God walks.[1]

Peace has a pace. A steady, resilient pace. And sometimes you have to slow down to recover your peace and learn how to move with it. Its pace may make no sense. You might be tempted to move fast because you know you can. But good work happens in the slow breaths, in the unrushed rhythms, in the spirit that swims against the streams of hurry.

To attain that peaceful pace, we must be willing to lay down the hurried pace we set for ourselves in the first place—the speed we thought we needed to maintain to keep up with everyone else. I think women especially often feel an immense pressure to have it all and then maintain it all just because someone, somewhere, said it might be possible. But the more I unplug—and find the most unexpected joy flooding in—the more I wonder if I even want it all. Is it worth it to have it all if I have to be an exhausted and depleted version of myself to get it? I'd rather take the "all" that I have—the peace that's coming from setting a more realistic pace for myself—and pour it back into the people and plans that actually matter to me. Peace is a better reward than applause.

I think we're all a bit tired of racing around in a world that always shouts at us to "be more." Maybe we could learn to just *be* instead.

The pace that it takes to try to hold everything together at all times isn't healthy or sustainable—and it hasn't been for a long time. There will always be another hoop or hurdle if we keep going in that direction. Another mountain to climb. Another thing to prove. There will always be the temptation to accelerate to a breaking point.

But it isn't too late.

It's not too late to learn to move at a different pace.

Are you finally ready to slow it all down?

chapter twenty-eight

THE YEAR OF THE VEGETABLES

After I reluctantly made peace with our pile of Christmas cards, we tiptoed softly into the new year, almost like we were trying not to wake it up.

This is not my default posture by any means. I typically start each new year with a list of things that will somehow be different about me promptly after the clock strikes twelve, as if the new year were a fairy godmother and—poof—my desire for gluten would suddenly be replaced with a passion for running and an intense love of green juice.

But this year, between the health issues with our daughter and the unplugged hours, I found those layers of striving and hustling peeling off me and exposing something new. I decided to set no resolutions. If anything, I un-resolutioned myself from the pressure to be any bit different once midnight hit.

Our New Year's celebration was intimate. We shared a meal

with friends we don't see often. My childhood friend Lauren cooked seared salmon, butternut squash, and green beans garnished with slivered almonds. We caught up. We laughed. We shared stories and played games. As we passed around plates of green beans and butternut squash, Lauren asked the five of us seated around the table, "What will be your word for the year ahead?" This is a long-standing tradition between Lauren and me. We each pick a word to represent what we hope the year to come will look like.

Lauren always picks solid, stable, calming words to guide her. I envy that about her. I, on the other hand, usually opt for more intense words that make me want to overhaul my entire existence in approximately twelve months. My word for last year was *prepared*. I learned my lesson after choosing that word and would like a redo. I think I'm finally ready to put the intense words to rest.

We went around the table, taking turns talking about our words, but it was hard for me to focus because I was so transfixed by how good the green beans tasted. I didn't even know I liked green beans until I tasted these. They were salty and tender and perfect.

Everyone around the table turned to me. I set down my fork.

"I'm hesitant to choose a word. So many parts of this last year felt hard, and I don't want to pick a word that could possibly make things any harder," I said, half joking but also deadly serious. "Honestly, these green beans are all I can think about right now. I need the recipe. I think my word for the year will be *vegetables*," I concluded with a laugh.

"That's a great word!" Lauren cheered me on. She went into a quip about how the nature of the word *vegetables* was bound to teach me something about growth—about the slowness and beauty of the process that takes place when something that was once buried in the ground begins to emerge and bloom and take on a new form.

I floated into midnight, imagining a quiet year of bok choy and turnips. Visions of eggplant and tomatoes danced in my head. I saw myself standing at the oven sautéing carrots and cabbage, or strolling the aisles of the local farmers market picking up bundles of kale and fresh herbs.

My veggie-filled visions came with a simple prayer: "Dear God, I'm not planning to move mountains this year. If you need me, I'll just be over here eating my Brussels sprouts."

———

I moved into the new year chopping, steaming, and sautéing. Within the first few weeks, I learned that to embrace vegetables, I also needed to embrace slowness. The two go hand in hand.

But we're so hurried, though. Who has that kind of time to prepare things? We drink coffee on empty stomachs. We rush through life. We don't have a minute to pause, never mind mince. Our schedules are full. There are bills to pay, kids to feed, and places to commute to.

During the pandemic, friends and I talked about finding the silver linings in all the hard realities of COVID-19. A common thread emerged: We discovered that our lives had been full before the pandemic—but they were full of the wrong things. When the world shut down, some of us had the chance to ask ourselves, "What will I not pick up again when normal life floods back?" Many of us said the hurriedness, the need to cram our schedules full.

It's years later now. Somehow, the hurry is back, and our grip on it feels tighter than before.

I don't know how to make a new path forward, so I learn to

chop kale instead. I get into a rhythm where I come downstairs in the middle of the morning and prepare the veggies. I start getting up earlier so I have the time to go slowly. I chop the spinach. I halve the juicy cherry tomatoes. I slice red bell peppers and watch them sizzle in the pan. I crack a few eggs and sprinkle goat cheese over the top. And when the dish is ready, it feels wrong to eat it too quickly. I sit at the table. I slow down long enough to savor what I've created.

I read a few pages of a book. I sip my coffee slowly. I marvel at how many years I told myself, "There's no time for this." No time to take care of myself. No time to nurture myself. No time to slow down long enough to ask myself, "How can I fill my tank better so that I can show up more fully to my life?"

I plan more meals. I use the cookbooks that have been sitting in our pantry for years, begging someone to dust them off and flip through their glorious, glossy pages.

I'm surprised to find that there is an actual spiritual discipline called "slowing." Author John Ortberg defines slowing as "cultivating patience by deliberately choosing to place ourselves in positions where we simply have to wait."[1] Ortberg suggests learning to practice slowing by driving the speed limit, choosing not to open your phone while waiting for an appointment, and deliberately standing undistracted in a long checkout line.

I've heard stories about slowing down for many years. The thought was alluring and quaint, but it felt like it was for other people—or maybe for when life would magically slow down all on its own.

We trick ourselves into believing we'll reach a day when our to-do list stops growing and events stop being added to the calendar. *Let's get past this season, then we'll slow down. Let's finish this project, then the stress will go away. Let's do this once more, then we'll evaluate.* But that day never comes, does it?

Slowing down is not a grace that magically appears; it's a decision we make when life keeps speeding up. We must be the ones to pump the brakes.

The more I slow down, the better life seems to get. The hours stretch out as my hurried spirit tries to listen to the wiser spirit within me, the spirit that knows the truth: *You have enough time. Don't believe the lie that you don't have enough time.* I recently heard someone remark that saying we don't have enough time is really just an excuse. Instead, we should say, "This just isn't a priority for me right now," and see how that feels. It might shift some things into perspective and force us to reevaluate where we've been spending our time and energy.

I don't have time to make a nutritious meal, but I had time to listen to that podcast.

I don't have time to move my body, but I had time to scroll for the last twenty minutes.

I don't have time to schedule that doctor's appointment, but I had time to open all those promo emails.

So much of life—the good, good stuff—is going to take extra time. We can neglect it, or we can step fully into it. Each moment is a choice. We can dedicate ourselves to the winding, imperfect long haul and watch in amazement as a new kind of life—a slower, more restful kind of life—unfolds before us.

———

My friend Lauren was right. Vegetables are the best kind of teacher.

They represent slow yet solid growth. They represent patience in the wait for the harvest, and joy when it comes. They represent the art of savoring, the goodness of slowness. Amen to filling our plates with these things.

But I think they also teach us that sometimes, to access their fullness, we have to be willing to dig deep and pull things up by the roots.

I spent years being an eliminator of things. I told myself I thrived on restriction, rules, and diets—off an invisible cycle of shame and deprivation that kept me wanting foods I "couldn't have." But as I started to cook more and experiment with colors and nutrients, I found that I wasn't depriving myself in the process. I was having fun with the ingredients. I was adding so many nourishing things to my plate and I felt so much peace, joy, and freedom in learning to give my body what it needed, rather than withholding certain foods from it. I could finally hear the story my body had been trying to tell me for a long time, about the cycle of exhaustion and inflammation it wanted to break me from.

I know it sounds a bit "woo-woo," but our bodies hold so much inside them—so much stress and anxiety and trauma—and I think they try to communicate to us what they need. As I daily embraced slowness and began weaving it into the fabric of my life, I learned something deeper was going on.

The vegetables helped me key into the exhaustion I had been carrying for years—the socially acceptable exhaustion many of us are carrying. I thought burnout had to look and feel a certain way—that it would be completely obvious if someone were experiencing it—but it doesn't. It takes on many shapes and forms. It can hit quickly, or it can build slowly over time. Through scaling back and having less noise coming in, I clued into what my body had been trying to warn me about: *You think the key is pushing harder, but you're slowly burning out. You need to switch gears to become healthy again.*

I visited a functional medicine doctor. I had labs drawn. A few weeks later, I had answers for many of the symptoms of exhaustion I'd been experiencing: thyroid issues, imbalanced hormones, adrenal fatigue, and insulin resistance. My body had been trying to tell me things were wrong for a long time, but I'd only responded by telling it to get smaller.

Now, with names for the underlying problems causing my symptoms, the healing could begin.

———

When I first started writing about the veggies, I thought it would be a little story about slowing and how I've learned to embrace it more in daily life. Looking closer, I've realized it's also a story about healing and how the two go together in so many parts of life. So much of my healing has depended on slowing down—on finally embracing the brake pedal.

I wish I could say that mental, physical, and spiritual healing were more instant, like many other things in our society. I wish I could promise they would be delivered between 5:00 and 10:00 p.m. today, so long as you remember to click the "same-day delivery" button. It's ironic to think about the level of efficiency that exists in certain parts of our lives when there are still so many things about the human experience that will always be slow, unsteady, and marked by the need for waiting.

I have a notebook in which I write down everything I'm discovering in this healing process. Sometimes the words are hopeful, and sometimes the words express frustration because I wish the pace would pick up. I once imagined I would finish the notebook and then I'd be healed, but it's starting to look like there will be a second notebook. I spent years of my life in a state of burnout. Walking the road back toward health after running on empty is a painfully slow journey, but there's beautiful scenery to take in along the way.

So I keep looking up. I keep looking around. I keep looking inward.

For now, I'm in the middle of the story. I pull a cookbook from the shelf and lay it open on the counter. I spread the ingredients on the cutting board and pull the spices from the cabinet. I pluck parsley and basil from the small herb cart on our back deck.

I dice. I chop. I turn up the heat and pour oil into the pan. I listen for the crack, the sizzle. I let the grease and juices fly up and stain the pages.

Days and months from now, I'll flip back to these blotchy pages and see the proof. All the proof I need to remember something good, something vital, happened here.

chapter twenty-nine

BEFORE THE NOISE GOT IN

I've been talking with many artists, writers, speakers, and poets these days, and they all say the same things: "I feel weary." "I can't create things like I used to because I'm trying to make sure people like the things I'm making." "The process is no longer free and whimsical like it used to be. It's a constant hamster wheel, driven by other people's praise or lack thereof."

I know that space well. I was living there—a weary and exhausted tenant—when I first committed to the unplugged hours. My creative energy had been slowly draining away over the last few years. I was creating less; I was consuming more. I was hyperaware of what others were creating but felt parched in my own creative process. I felt like I couldn't make beautiful things anymore.

I sensed my spirit telling me that enough was enough. Enough filling my brain with information it didn't even know what to do with. I needed to fall back in love with the patient process of making things out of nothing—with the secret, deep work that lights the fuse for creativity to burn.

———

Something has shifted in our culture. It feels like we're doing far less secret work, taking far less time to explore our creative minds to see what lives there. We think we don't have time for that sort of thing. We're distracted and overstimulated so often.

In her book *Bored and Brilliant*, journalist Manoush Zomorodi writes, "Creativity—no matter how you define or apply it—needs a push, and boredom, which allows new and different connections to form in our brain, is a most effective muse."[1] The problem is, as my friend Christina once said it best, "We're not letting ourselves get bored anymore, so we're not being creative."

I believe that we're all wired to be creative. Different people have strengths in different areas, but I think creativity and imagination exist within all of us. Some of us write. Some of us paint. Some of us make meals. Some of us create deep traditions. Some of us design beautiful spaces. We're all creators who have, deep within us, the power to make. To invent. To design. To compose. To cook. To decorate. To script. To bring new things to life.

Years ago, I got a small tin box and placed it on my desk. Whenever I needed to do deep, focused, creative work, I put my phone in that box and wouldn't remove it until I'd made something. That was my first small step into unplugging, and it became my rhythm for a long time: no consuming until I'd created something. But I found myself getting away from that ritual as the urge to mindlessly consume took over and new apps kept popping up to make overconsumption even more readily available. The more I opted for consuming over creating, the more my creative spirit was worn down.

I can't say digital technology is entirely responsible for stealing away our creativity and imagination. That wouldn't be fair. But I've also seen how we can lose minutes and hours by clicking and swiping. I've seen how consuming video after video and post after post can dull the parts of us that ever wanted to create things in the first place. Creativity is a muscle. Muscles must be trained and flexed and built and repaired. The muscle of creativity doesn't grow from watching what others have created. It grows from practice. It grows from letting our minds wander off. It grows from taking time to breathe and think and take in the world around us. It grows from experimentation and the willingness to dig deep enough to be original when it would be easier, in today's copy-and-paste culture, to imitate what everyone else is doing.

I think we can thrive on a certain level and certain types of consumption: Documentaries and beautiful films can inspire us. Relevant news can push us toward action. Classic literature can make us think deeply. Good meals, well, good meals make almost everything better. These things can fill us up and be regenerative, but there's a tipping point.

I'd say we've lost the right balance. We consume at all hours of the day—early in the morning when we first wake up and late into the evening, even though scientists warn us about all that blue light. We consume without knowing why we're consuming. We consume without realizing the consequences—that everything we take in, every image and video, impacts our souls. Each fifteen-second clip or three-minute video is another thing that can alter how we think, feel, and interact with the world around us. When our attention leaves our present moment, we step out of the natural, creative flow of life and become less inspired. We lose touch with the parts of us that feel more alive when they're working with raw materials: wood, herbs, numbers, sentences, dough, sweat, or musical chords. We forget we were made to make something out of nothing.

————

When I was younger, I used to write cryptic novels that would now find a perfect home in the *Gone Girl* era of literature. They were always about missing children and grief and murder. My teachers kept recommending guidance counselors, but my mom just let me keep writing. One of my most notable novels, which strangely never won awards, was a hybrid between the stories of Aladdin and Anne Frank. I wrote it when I was eleven.

A novel a year. That was the calendar by which my adolescent life operated. At the end of the year, my mom would take me to the public library, and we'd print out dozens of copies of my latest novel and staple the pages together so I could gift all my relatives the latest release on Christmas Eve. May God bless my relatives; that's all I can say. And may God bless my guidance counselors.

I eventually stopped writing those novels.

It wasn't because that earnest creative soul vanished in the middle of the night one night. No, losing her wasn't instant. She slipped away little by little, bit by bit. The noise got in. All kinds of it. The noise of other people. The commentary. The likes and the approval. And then there was idleness. Then overconsumption. Then the need to be praised, followed by the need to be loved.

Soon enough came the desire to be successful. And then the need to prove myself. And then the intoxicating sound of applause. Before long, that girl I used to know wasn't writing imaginative stories anymore. She wasn't making up worlds in her mind. She wasn't finding inspiration at every turn. She no longer knew how to operate if people weren't cheering her on.

The noise got too loud. And then the noise was suddenly everywhere. Everywhere, everywhere. And that creative girl was nowhere to be found.

Maybe something similar happened to you. You used to have a certain part of yourself that you loved so much, that felt special and uniquely you, but time passed, and you looked at yourself in the mirror one morning and thought: *What happened? Where did that part of me go?*

So tell me the truth: Who were you before all the noise got in? Who was that person?

I want to know all about that person. Tell me the stories.

Were you funny and brave? Did you like to watch old movies? Did you do things with your hands or write words deep into the night? Did you cook? Did you sew things? Did you pray more? Did you go for long runs where it was just you and your feet pounding against the pavement?

One friend told me that when she was little, she used to bake all the time. Her childhood was marked by flour, eggs, and mixing bowls; she told me that baking stirred a deep satisfaction within her. Now she doesn't bake at all. What was it, before all the noise got in, that made you come alive?

I often jot down tiny notes for my daughter, keeping track of what she's loving. Cars and princesses. Making art. Right now, she's entranced by the idea that the entire world is a stage. She makes us sit at the dining table with our chairs facing her, and she dances around the room, holding a thick washable marker as a microphone. She is full of joy and delight in those moments. She is free.

I keep these notes because she may reach a time when she forgets who she is, loses herself to the noise, or can't find her passion any longer. These notes may not hold all the answers, but they may contain some. Maybe she will find herself tracing these notes and remembering: *Oh yes, I loved to make art before all the opinions got in. I loved to dance before the praise became too much. I loved to sit and read a book, constructing massive worlds in my head, before I became so consumed with what others thought of me.* She may need help remembering at some point. And if she needs reminders of who she was in these early years, I'll have the breadcrumbs waiting for her.

———

Lately, when people ask me how the unplugged hours are going, I tell them this: I'm getting back to the person who was there before all the noise got in, and I didn't realize how much I missed her.

That person, she loved art and decorating. Whole worlds existed in her brain. That person loved creating for the sake of creating—even if the words would never be shared, the art would never be displayed, and the book would never be published.

That person is coming back to life. And it didn't come through carving out hours to find her or through picking up a new hobby. That wasn't it. I turned down the noise and kept turning it down. I put myself in places where my mind was free to wander and think deeply rather than grab for the next bite-sized piece of content popping up on the screen. Eventually, the girl who existed before all the noise got in began to speak again.

The other day, I finished work at a coffee shop and hopped in the car to drive to an appointment. I was sitting at a light, waiting for it to turn green, when a conversation popped into my brain, seemingly from out of nowhere.

I knew instantly, in a way maybe only other writers will understand, that this dialogue was between two characters in a novel I've wanted to write for the last thirteen years. It's funny because thirteen years ago is exactly when the noise started getting in—the same year I got a smartphone and started a blog that opened the door for both praise and criticism. I instantly knew the sounds of the two characters' voices and marveled at how the conversation took shape in my mind, almost like I was watching a movie scene flickering out of a projector.

Flustered by the sudden creative surge, I pulled into the nearest parking lot, yanked my computer out of my bag, and typed on the keys furiously until I got the whole dialogue down on the page. Out of nowhere, I was ugly crying in a way that must have made me look like a mad woman in the middle of the Target parking lot. I was surprised and terrified by the whole moment.

Here's the best way I can describe how it felt: It was like hearing a song you used to know by heart after not hearing it for many years. The song held deep meaning for you—it marked an important time in your life. Years passed, and you'd all but forgotten it entirely until one day, it streamed on the radio out of the blue. The words hit you, and before you knew it, you were sobbing and remembering all the significance the song carried. You never thought you'd forget that part of you, but as it turns out, you did.

It felt like coming home to myself, like I'd pulled up outside and found that a part of me I'd forgotten about was standing on the front porch waiting. And when I got close enough, I heard her say, "You're back. You're finally back."

That's happening more and more these days. Bits of novels are flooding into my mind every time I power down. I catch them in the yellow notepads I've planted all over the house, and then I move on with the rest of my day. It feels natural, but it also feels so inspiring to hear myself above the noise—like I'm getting back pieces of myself. I love how my friend Nigel phrased it to me after commenting on our many years of letting too much noise get in: "Now is a great time to remember who we are at our core."

It's never too late to remember. There is still time to recover who you were before all the noise got in.

chapter thirty

AVAILABILITY

Every year, I look forward to a December tradition—one that has withstood the tests of time and circumstances, moving boxes and life changes.

It began over a decade ago in New York City—a group of girlfriends would get together every New Year's Eve to make vision boards. We'd spread glossy magazines across the floor and spend hours chatting about our dreams, flipping through the pages for inspiration, ripping out phrases and images that spoke deeply to us. We'd cut and paste and concoct vision boards for the year ahead.

Over time, that group of girlfriends dispersed. Some left the city for good. Some moved to different boroughs. We were no longer able to meet in the same space every year to carry on our tradition. So, for a few years, the tradition stood still.

During the pandemic, one of the girlfriends emailed the rest of us and said, "I need this tradition back in my life."

We all agreed that it was a year when we needed one another, and a little extra vision couldn't hurt. We picked a date. We met up over Zoom. And from our various corners of the country, we flipped through magazines and ripped out phrases and images that spoke deeply to us. We carried our fresh visions into the year ahead.

We've stayed loyal to our Zoom call for the last few years. This tradition of ours is a perfect example of the nuance involved in decisions about technology. Without tech, we couldn't connect with one another year after year in this way. So I keep telling myself the same thing: The tech is good so long as we're intentional with it.

In our most recent Zoom fest, I found myself creating a vision board that felt oddly different from my boards of years past. Color features prominently on this latest board—which typically never happens. And the board contains images, which is also a shocker—I'm used to picking out big, loud words to plant all over my board so I can move into the new year accosted by magazine cutouts of directives like "RUN" and "GET MORE DONE."

This new board is softer. Scattered across it are images of puzzles and people sitting around tables. There is food. There is decor. There is gathering. There is life being made. I picked a few good words to place around the board: *Cozy. Magic. Live Lighter. Wide Awake. Festive. Something New.* Yes, this is entirely something new.

Glancing at my board that evening, surrounded by glue sticks and stray magazine pages, I felt like the board was telling the truest story I'd told myself in a long time. I'm finally making room for what matters most to me. It's not the hurried pace. It's not the frantic goals. It's not money or success. It's people. And it's good meals. It's presence, peace, and the chance to bring on more joy. I'm here for all parts of this new vision—a vision that birthed itself through many unseen hours and the earnest attempt to practice presence day after day.

For years, I prided myself on being busy, which was evident from just a glance at my vision boards. They say being busy isn't a badge of honor, but I wore it like one until the stitches fell out.

I didn't just want to *be* busy; I wanted people to know I was busy. This was tied to that raging insecurity within me—the fear that I was never doing enough. When friends asked how things were going in life, my response was always the same: "I'm just so busy."

I'm busy this month.

I can't make it this week—I'm busy.

Did I mention that I'm busy?

The more I power down, though, the more I find myself craving whatever is the opposite of that busyness. I've started calling that opposite "availability."

For years, I confused availability with accessibility. We've tied our concept of availability to the idea of constant connectedness. We assume people should always be available because of all the digital devices we possess. We get anxious when we don't hear back almost immediately. *Are they mad at me? Did I say something wrong?*

I don't think we're called to an availability that robs us of peace or burns us out for the sake of always being accessible. I don't think we're called to be yes machines—always willing to add another committee, task, or engagement. In my mind, availability is less about being "on" for everyone all the time and more about making space—for people to be seen, for needs to be met, and for our lives to flourish as we take the time to care for ourselves.

There's a moment at the beginning of the book of John that I often return to when I consider the idea of availability. It's the story where John the Baptist stands with a few of his disciples and sees Jesus walking by in the distance. John calls out, "Behold, the Lamb of God!" (John 1:36 ESV). I always wonder how that declaration was received. Did the men with John think he was crazy, or was this just casual John on a Tuesday?

Two of John's disciples are curious about this Jesus character. They approach Jesus and ask, "Where are you staying?"

"Come and you will see," he replies (John 1:38–39 ESV). The two men follow him.

This is one of many places in the biblical text that show just how different Jesus and I are. "Come and you will see" is not even remotely part of my vocabulary. I'm more of a "Come and see on Friday at 6:00 p.m. because that's what we've planned" or "Come and see three weeks from now when my house is picked up" type of person. I get instant anxiety imagining these men following Jesus to wherever he was living. I at least want to run ahead and place the coffee cups in the sink for him and make sure the bed is made.

But the story doesn't get less anxiety-inducing. There's more. The text says the two men stayed with him until the tenth hour, which would have been about 4:00 p.m. Translation: They spent the entire day together.

The planner in me, who likes to precariously divide her day into fifteen-minute increments, cringes at the thought of an entirely impromptu activity taking up the entire day. I hope Jesus at least offered them snacks at some point. But here is me, missing the point, and here is Jesus, making himself available to two men clearly searching for something.

I imagine he simply invited them into the rhythms of his daily life. I bet he asked them questions and got to know them better. He extended his availability to them—his undivided attention.

———

I've learned a lot about this kind of availability from my friend Cara. For as long as I've known Cara, she has been showing up at my front door. Randomly. Always unexpectedly. If I ever open the front door to find a gift, book, or random treat, I can be assured that Cara's car was parked in front of the house for a few moments.

I know dozens of others out there are nodding in agreement because they either know Cara or have their own Cara in their lives, and they get it.

It started with a tin of cinnamon rolls dropped off on our doorstep one December, and it hasn't stopped since. Every year, a small box of cinnamon rolls appears on the doorstep just days before Christmas. And I can guarantee you that Cara is driving all over Atlanta, dropping off similar tins on other doorsteps.

From Christmas cinnamon rolls it morphed into meals and thoughtful care packages. Gifts and "just thinking of you" messages.

Last year I had to travel to Baltimore for work, and I felt anxiety forming in my chest because it was my first speaking engagement away from home since having a baby and surviving a pandemic. Cara recruited another girlfriend, tracked down the hotel where I was staying, and arranged for a gift to be delivered to the front desk for me. She stopped everything she was doing to send me a message: "I know this is a hard trip for you, but you're not alone. We're with you."

Cara recognizes the moments we're tempted to overlook or downplay and makes sure we take a collective pause to see the bigness in even our small victories.

But here's my favorite part: During Cara's baked-good drive-bys that happen throughout the year, it's almost guaranteed that one or two of her little boys are buckled in the backseat, staring out the window as she runs to the door, lightly knocks, and then jumps back in the car. They're along for the ride. They're watching her. They're witnessing her love in action. And I know, without a doubt, they are marked daily by the way she uses her availability to make others feel known and loved.

Cara has taught me with her own life: Instead of always waiting for someone to show up at your door, you go first.

I've said for years that I want to be like Cara—to be the kind of person who shows up in that way—but I did nothing about it. I kept filling my calendar. I kept frantically moving from event to event. I kept struggling to remember people's birthdays. I kept living a life with no margin—a life with plenty of accessibility but very little true availability.

A recent trip to visit my mom in Connecticut was the tipping point for me. My mom's birthday fell on the last day of our trip, and that whole day, people kept coming to the door. Each time my mom opened the door, it was another friend. One came with a glass trifle bowl full of shortcakes, freshly cut strawberries, and homemade coconut whipped cream. Another friend entered the house with a Whole Foods bag packed to the top with brie cheese, macarons, chocolate, flowers, and dried fruit. The next-door neighbor walked over with a gift bag from his wife, packed with some of my mom's favorite things.

The intentionality behind these gifts stirred up something deep within me. I told anyone who would listen about this mystical experience. For weeks, I couldn't stop thinking about all those people coming to the door—taking time out of their day to bless my mom, not because she was in crisis mode but simply because she was worth showing care for.

When the thought of these people's availability and generosity would not stop haunting me, I intentionally made a new priority: We will be the people who show up at the door. And that declaration has morphed into one of the most life-giving traditions for my daughter and me. Once a week, we find a space in the calendar and use it to see someone else, to make them feel known and loved.

We go all out for the occasion. We fling open her closet doors and pick a frilly, over-the-top princess dress for her to wear out. We pair it with plastic heels or combat boots and we either visit the local coffee shop or sit in a booth at Waffle House and feast on chocolate chip waffles, cheese eggs, and endless refills of black coffee for me. We weave a shopping cart through Target or the supermarket and pick out a few little things for the person we've chosen to bless. We put together a gift for them and show up at their door (or sometimes their workplace) to say: We love you. We're in your corner.

It's not necessarily an ambitious or extravagant gift, but no matter the size of our act, we make it the centerpiece of our day. We treat it like it's our most important mission—because it more than likely will be. Sometimes we draw pictures for friends and drop them in the mailbox. We make cookies for Dad. We drop flowers at our neighbor's doorstep. We deliver homemade baked goods to friends across town. As often as we can, we show up at someone's door to tell them, in the flesh, "We're rooting for you today."

Novalee is too little to fully understand what we're doing yet, but I think one day she will. The childhood memories that stick in my mind are all the times my mom showed me how to make the intentional choice to bless someone else with my time, with my gifts, with the little or the much that I had. When Novalee and I began our weekly tradition, I thought we were clearing the space to bless others. In actuality, the practice blesses us. It's regenerative. I feel lighter and more joy-filled as we take our eagerness to bless others and turn it into a lifestyle—one small act of love at a time.

So many good, unshakable things have compounded over my year of unplugged hours, but this is one of the best. And it's a practice I recommend to anyone who feels like their life is overrun by the wrong things—by the constant streams of texts, by the mindless scrolling, by the things on your calendar you said yes to only to please other people. You will never regret clearing the space to bless someone when they're not expecting it. You will never regret showing up at the door.

As I write these words, my vision board is hanging on the set of antique lockers beside me. From the corner of my eye, I can see the colors popping. I see traces of the divine woven throughout the words and pictures.

There's a recent addition in the middle of the board. An addendum, if you will. Scripted in black Sharpie on a plain index card are the words *Clear the space*. These words say it all. They bring the vision together and hold all the good parts in place.

chapter thirty-one

SPACIOUS PLACES

My daughter's love for princess dresses runs deep. It's deeper than any obsession her two-year-old spirit has ever latched onto before. She's *in it*. For her, every day is punctuated by outfit changes. I watch her shimmy into a purple flowered dress and run over to the mirror in the next room to take a look. Her whole demeanor changes in a moment. Suddenly, she's regal. She's confident, and she's joyous. She takes her own breath away. A massive grin spreads across her face as she turns her small body ever so slightly to see the back of the dress and then speaks out loud to the mirror, "Look, how beautiful."

She turns back to see that I'm watching.

I'm in awe of her. She carries a confidence that tells me the world hasn't gotten to her yet. She hasn't yet grappled with comparison or insecurity. She lives in a spacious place, untouched by what other people might think of her. Untouched by the voices of fear, hurry, and anxiety. I want to keep her in this space for as long as possible.

Sometimes I watch her with a tinge of envy because I'd like to be that free. But maybe I was once, and maybe that freedom isn't completely lost to me. Perhaps there was a time when I knew that spacious place too, when I felt it deep within me and operated from that space of vibrant overflow.

————

Something within me is dying lately. It's my need, my constant desire, to "fix myself." I'm doing everything in my power to try to keep it on life support, because I don't know who I'll be if I'm not trying to fix something broken within me. I've been a self-improvement junkie since the age of twelve. My whole identity has been built on thirty-day challenges.

It feels like whenever I get online, I'm bombarded with videos on how to fix myself. The algorithm has really ramped them up lately. I can just imagine whoever manages the algorithm scrambling in the back room, saying, "She's been gone too long! Time to pull out the big guns! Show her a new diet! Give her a twenty-one-day challenge! Let's reel her back in!"

But something in my spirit is more cautious now. It's saying no more. Not another diet. Not another program. Not another software. Not another app. Not another book written by another expert. I was made for more than this. I won't carry this yoke of constant self-improvement anymore. This ends now.

I can't remember a time in my life when I didn't treat myself like a project. Even when I started tracking the unplugged hours, that's what it was to me—a self-improvement project. There was a pivotal moment around the eight-hundred-hour mark when I stared down at my paper tracker—tattered and ripping on the edges—and realized I would hit my goal: one thousand hours unplugged in one year. I'd been diligently tracking every unplugged hour since I started.

I stepped back and said to myself, "You're going to complete this challenge. Another thing checked off the list. But what if you hadn't hit the mark? Would what you've already done be enough for you? At what point do you decide that it's enough and you're proud?"

Each tiny bubble filled in on the page represented a gift to myself—a little nugget of time given back to me. Seeing them all sitting there, so close to the end of what I'd set out to do, I couldn't deny I'd become a different person in the process.

I resolved right there, at eight hundred hours, to stop seeing the challenge as pass or fail and just be proud of the progress.

I'd be proud of the consistency, not just the goal met.

I'd be proud of the continuing to show up, not just the result.

That resolution is one of the bigger gifts to come out of unplugging, and it's one I wish I'd latched onto sooner: the idea that we're not projects. Our lives don't boil down to self-improvement tactics or personal growth challenges or a dream weight. We're meant for more than striving to become some ideal self who may or may not finally make an appearance. We're meant for more than continually "waiting for life to begin" because we don't yet feel worthy of the life in front of us. I know we're meant for a deeper, more soul-filled way of living.

I think we get to decide, once and for all, that we are proud in this moment—proud of how far we've come and how much we've grown in the process. No matter what, we're going to be proud. We can release our grip on the ideal we've built in our minds and decide to love what's already here. That doesn't mean we won't progress or become new versions of ourselves along the way. It simply means we agree to stop withholding goodness from ourselves, thinking we'll only deserve it when we meet some impossible set of standards. We're learning every day. We're growing every day. We're showing up every day, even though life feels hard and out of control so much of the time. It's time to be proud in the process.

We were never meant to reduce our lives down to a series of check marks and like buttons or to try to fit within boxes that are too cramped for us. We were born for more than productivity trackers, five-step programs, and diets that keep us in a constant state of hunger.

We were meant to be free—to live in spacious places. We were meant to look around us with the same grin my daughter wears on her face when she looks at her reflection staring back in the mirror and remarks, "Look, how beautiful."

I want to believe that is for us.

I've been getting more comfortable with sitting in the stillness. I'm actually starting to look forward to it—that sliver of time in the day when my phone is powered down and I can sit still. It feels like I'm still in the middle of a grand deprogramming from my years of needing to occupy my time with betterment and improvement. Instead, I sit and wait. I listen.

In the silence, one scene keeps coming to mind.

It's a story from the Scriptures, one I've heard referenced in many sermons over the years. It's the moment, just before Jesus's public ministry begins, when he is baptized. As he emerges from the water, a voice comes from heaven, saying, "Thou art my beloved Son, in whom I am well pleased" (Mark 1:11 KJV).

I can still hear the preacher's voice saying, "Before he ever did a thing, he was loved." I've always thought, *That's a great,* comforting statement for people who need it, but I don't think it could ever be that simple.

The more I push away from all the noise in my life, the more I wonder if those words could be for me too. Henri Nouwen once wrote about a voice that calls us beloved without our needing to prove it.[1] This kind of love doesn't hinge on performance. It doesn't require an accolade for admission. It doesn't demand constant improvement out of me—none of that.

Something is shifting inside me. Something is growing as I enter the stillness and learn to listen for the voice that sounds like love. Like favor. Like pure delight in who I am, exactly as I am, in this moment.

I'm learning to distinguish that voice and follow its call; it feels like moving through a deep patch of woods and unexpectedly stepping out into a clearing. Into a spacious place. I didn't see the spacious place coming on the map. I didn't even know it was there.

But as I step into the clearing, I'm met with a familiar peace and a feeling that I've been here before. A subtle grin spreads slowly across my face as I hear my own voice saying out loud, "Look, how beautiful."

chapter thirty-two

PROCEED QUIETLY

The day I reached the one-thousand-hour mark was anti-climactic. But looking back, I think that's maybe the point.

I knew the milestone was coming. We were approaching the one-year mark as we returned to that same beach house with friends where the story had begun—to the same screened-in porch where I first heard the gentle nudge in my spirit say, *Turn off your phone.* When I powered down that first time, I had no idea I wouldn't just be taking back my time; I'd be taking back my life.

But if you were to ask me if there was a specific moment when I crossed off that final hour or even did a celebratory dance or victory march, I'd have to be honest and tell you no. My phone was off for nearly that entire trip. The unplugged hours reached the one-thousand-hour milestone and moved beyond it rather quickly.

Somewhere in the middle of the trip, I remember sitting on a pool float in the mid-morning sun. My daughter was napping. My friend and I were reading novels—floating past one another and breaking the silence every few minutes to update each other on the plots unfolding.

I was completely undistracted and fully enjoying myself. And I realized, as I sat there with my book, that I had completed the challenge.

The moment wasn't big. I didn't feel the need to capture it in some concrete way. I didn't need fireworks to go off in the distance or a marching band to come tromping through the backyard—none of that.

The moment felt simple yet profound; it was all mine. It would have been easy to share the news as fast as I could—to rake in the applause that used to be my fuel. But I knew the gravity of all that had happened over the last year. I knew it, and I felt it. Those who love me felt it too. And for once, I didn't need anyone else to tell me it was valuable or good enough. I knew it for myself and wanted to sit in that feeling for as long as possible.

I paused. I waited. I chose to savor.

And in the savoring, my spirit was satisfied.

———

There were certain questions I repeatedly asked myself as I ventured into my initial unplugged hours: *Does this moment still matter to you if no one else knows you went, saw, lived, ate, loved, fought, and tried? Does this moment still matter to you if you never pull out your phone to tell people that it happened?*

At first, I didn't know the answers. I tried to convince myself it still mattered, but I wasn't sure. I'd lived the last ten years wondering, *If I don't share, did it happen?* It's a habit that can't be broken overnight.

I'm starting to think that the healthy opposite of constant sharing is learning to savor things for ourselves.

To savor something is to enjoy it slowly. To take it all in—maybe from multiple angles. It's focusing our awareness on the moment at hand, without distraction or hurry. It means rooting our heels deeply in the ground, daring to feel it all and not move from it too quickly. It's the practice of sitting with a sliver of time we'll never have again.

We'll never experience our life this exact way—in this current season, with these current circumstances—ever again. Savoring is accepting that the present moment is fleeting but choosing to lean in anyway.

———

Lately, there's a mantra playing in the background of my life: Proceed quietly. I hear it in nearly everything.

When I'm making a meal and I'm tempted to share the recipe online. *Proceed quietly.*

When I'm experiencing a moment that fills me with deep enjoyment and I want to sum it up in a grainy photo. *Proceed quietly.*

When I come across a gorgeously crafted sentence, and I want others to know it exists in the pages of that book. *Proceed quietly.*

When I've accomplished a task and I want that quick hit of dopamine that comes from sharing the accomplishment with a larger audience. *Proceed quietly.*

The voice feels like a peace-filled whisper—the opposite of the anxiety I listened to for so long—so I lean in to heed its wisdom every time.

I realize "proceed quietly" is not exactly the mantra of our world. The world tells us to make a ruckus. Make noise. Be the loudest. Let people in on what you're doing. Share, share, share.

Sharing is a vital part of life, but there's a difference between sharing from an overflow of emotion or as an afterthought and sharing because you believe your value and worth hinge on it. Many of us harbor an "I share, therefore I am" mentality, as sociologist Sherry Turkle would call it, that we need to shatter if we ever want to live fully present lives.[1]

Even as I sit here writing, my fingers tapping against the keys of my laptop, I hear the invitation pop up again: *Savor this deeply. All the evidence you need of a life well lived is right here: a candle burning; a plate scraped clean; a favorite book open, spine up, on the desk beside you; the sound of laughter coming from downstairs; sunlight pouring through the window; a phone turned off and tucked away. All the evidence is right here.*

Proceed quietly.

chapter thirty-three

TAKE YOUR TIME

When I reached the one thousandth unplugged hour, I imagined the story would be finished. I'd have completed a challenge, and it would be time to find another one to put in its place.

But here I am, standing well beyond the one-thousand-hour mark, and I can tell you a different truth than what I anticipated: This story is only beginning. There's so much more to live and understand, so much more to uncover and learn as the unplugged hours stack higher and higher. Every time I think I've learned all that I can from the unplugged hours, another lesson emerges, ready for me to go deeper—something new that I never saw within myself when the phone was buzzing and the emails were pouring in at all hours.

It's almost as if those first thousand hours set the foundation, and now I'm building upward—erecting frames, building walls, and dreaming about where all the furniture will go.

I'm a different person than I was when I turned off my phone for the first time. That time in my life was marked by a frenetic pace and the belief that there was never enough time.

"I don't have enough time," I said whenever I felt stressed or anxious, or when there was undoubtedly too much piled on my plate. But I've learned, through the steady rhythm of powering down, that the time was always there. All along, it was available—it was just waiting to be reclaimed. It was waiting for me to see that I had been putting it in the wrong places and had to learn to use it differently.

So that's what I did.

Since the unplugged hours began, there have been hours of presence and deep connection—interactions so rich and real, all of us there forgot to check our phones.

There have been hours of movement, of reading good books. Hours spent entertaining guests and cooking nourishing meals. There have been puzzles and walks and moments gathered around the table to drink life down deep with one another.

There have been hours of seeking God, falling back in love with the rhythms of spirituality, praying in secret spaces, living more curiously, and creating new things—at the table, at the counter, at the desk in the corner of the room.

Hour by hour, I came back to life. I rediscovered the color that had faded from the picture. It feels like everything came back, even things I didn't know were absent at first, but it all came back even better—like newer, upgraded models. Just a few minutes ago, as I was sitting here at my desk, I heard the front door swing open and closed. Curious, I left my chair and went downstairs to find Lane and Novalee sitting on the front porch, listening for crickets in the night air. Such a simple rhythm of peace that they savor together at least once a week. Years ago, I might have missed this.

Tonight, I went out onto the porch and stood with them in the moment—I dug my heels down and stayed right in it.

When I talk about time in the years to come, I hope I can say there was always enough of it. It wasn't too short, and it wasn't too fleeting. I didn't try to catch it or bottle it up—I dug my heels down, determined to experience each moment to the fullest. And I watched as my hurried language changed from "I don't have enough time" to "I've had the great gift of time—and it has been enough for me."

I don't need to prove that I use my time well anymore. I don't have to wring out every last drop for efficiency or productivity's sake. I don't even have to document all of it. I know the hours were good. I know they were sweet. I know because I was in every one of them. I was totally and completely present—whether it felt sweet or hard or holy or uncomfortable. I was present. And that made all the difference.

For the last few days, I've been wondering and praying about the words to leave you with. What words could I give you for the end of this story and, hopefully, the beginning of your own? There are so many travel notes I could bestow for the journey coming up ahead. So many words, but I've settled on three:

Take your time.

It's as simple and as difficult as that.

Technology will keep moving forward. It will only get faster and faster. There will always be new advances and new temptations to plunge deeper into digital connectivity. But you get to decide when enough is enough. You get to set the pace for your own life. You can slow things down. You can find your own rhythm. You can learn to sit in the silence and the stillness. You can approach your waking, breathing life with deliberate care and attention. You can take your time in a world hell-bent on convincing you there is never enough of it.

But I think those three words—*take your time*—run even deeper than that. They carry another meaning, a challenge I want to leave you with—a challenge to take your time back.

Take your time back within a culture that will always move frenetically, making you feel like you'll never get it all done. Take your time back for the better things of this lifetime—the things that rarely happen on a screen.

Take your time back for deeper connections with the people you love—the chance to get to know them all over again. Take your time back for all the beautiful side effects that come from learning to pay attention when it would be easier to divide that attention into tiny, disconnected fragments.

Take it back for dancing and for staying up too late into the night talking. Take it back for spontaneous adventures, for early-morning coffee dates, for chances to go see the holiday lights. Take it back for dinners at the table where the phone doesn't sit beside you, for making meals or cozying up on the couch with takeout, for dreaming and getting lost, for concerts in the park.

If you've been frantically saying to yourself, "I don't have enough time," then begin taking it back—one hour at a time. For enjoying lazy days and forging new hobbies, for falling in love with people and places and projects that matter to you. For carving out spaces free from the noise, for scrapping the sense of obligation that says you should always be in attendance and deciding to just stay back this time. Take it back for chances to show up at someone else's door. Take it back for going places without your phone. Take it back for all the things you always wanted to do, the things you stopped doing when the pace got too fast, the overload exhausted you, and your spirit became weary from too much noise. It's not too late to take it back. No, it's not too late.

Take your time.

Take your time.

Take your time.

ACKNOWLEDGMENTS

I learn anew every time I set out to write a book that the finished work is the product of a village—of many people who bring the words, lessons, anecdotes, and stories to life. Over the years of living and writing this message, it became more than evident: I have a beautiful village.

First and foremost, I'm thankful to God for the opportunity to write these words. In every fringe hour I could find, I made myself sit still and ask for the right words to fill these pages—and God met me every time in such a gentle, kind way. If you encounter more of the peace and presence of God as you read these words, then my work is done and done well.

To Laney—from the moment we first met, you've been my biggest supporter and champion. I could not do this work without you. Thank you for every pep talk, every breakfast and dinner that got me through the lengthy writing hours, every bath drawn, every cup of tea, every card written, every prayer spoken over me, and every sticky note planted around the house. You're my very best friend, and I'm the luckiest.

To Novalee—my darling girl, you served as the inspiration for so much within the pages of this book. Thank you for teaching me to take in the awe and wonder around me. Thank you for your sweet spirit and kind heart. In a world that is often dark, you are a spacious meadow. Never change, Kook.

To Mom and Dad—long before this book existed, you taught me the importance of living what you write down. You've been steady and faithful my whole life. Thank you. Mom—if anyone could speak to cultivating a life of deep presence, it's you. I love you.

To Baccu—I've written every book for you, but this would have been one of your favorites. It's been nearly two decades since I've seen your face, but I still feel you with me all the time. Your presence has staying power, and that's the heartbeat of this book. I love you and miss you.

To my family—thank you for always believing in me, for loving me so thoroughly, and for covering me in continual prayer all along the path to bringing this book into existence.

To Grace Anne—I never could have written this book without you. Thank you for being my sounding board, friend in doubt, editorial partner, and first reader of every sentence, draft, and rewrite. You are a marvel, and I can't imagine my life or work without your presence.

To Dawn, Tory, Hayley, Cara, and Dee Dee—within the unplugged year, your friendship became one of the most notable highlights, and my deep admiration and love for each of you have only grown since. Thank you for spurring me on and making me better. God gives the best gifts. You, girls, are my living proof.

To Christina—you've been the most faithful and beautiful friend. I've loved seeing our friendship move through so many notable seasons. Thank you for the ways you call me upward and remind me to take up space.

To Kami—I wasn't expecting our friendship, but I can't imagine my life without it now. Thank you for cheering me on in every "resistance task." I love you so.

Acknowledgments

To Elyse—from the ranch to the big couch, you've been a constant cheerleader throughout this writing process. God knew I needed your laughter, spice, and immense faith to enter my life at the exact moment they did. If my appreciation for you were a puzzle, it would easily be one thousand pieces.

To Lauren, Melissa, Rachel, and Tina—thank you for being my voice-memo prayer warriors on bright and hard days. You keep me going, and I am so indebted to you.

To Mackenzie—from the moment I brought this idea to you, you were my fiercest advocate and cheerleader. Even more than that, you've been a faithful and fierce friend. Thank you for teaching me to be brave in my writing career.

To Christine—your editorial guidance throughout this process was golden, and I am so thankful for your gentle nudges and belief in my message. Thank you for making me more confident in my writing style. You made this writing experience so complete.

To the team at Zondervan—thank you for believing in this book and working tirelessly to bring it into existence. Thank you for giving me a publishing home that makes me feel seen and known daily.

To Carolyn and Carly—thank you for championing this concept so fully. I could not have imagined this process without either of you!

To my church family at Passion City—thank you for your goodness and consistency in an often flighty world. Grant and Maggie—thank you for your sound leadership. Louie and Shelley—thank you for being pillars in my life and teaching me to walk faithfully, step by step.

To Taproom—another book written within the walls of your coffee shop. Thank you for creating a space where the coffee is strong, the baristas are kind, and the company is unmatched.

To my Groove community—thank you for cheering me on

throughout many early and late writing sessions. You helped me gracefully string together the words in this book—fifty minutes at a time.

Once again, there are too many people to thank—in Atlanta, Connecticut, and beyond—people who sat beside me, refilled my cup, read sentences, linked arms with me, lived out the concepts within these pages, kept vigil, and prayed big prayers. You know exactly who you are. Thank you for embodying the presence and benevolence of God so thoroughly in my life.

Last but never least—to my readers. You make this work so vibrant and fun. You have for the last decade. I thank God for you all the time. Thank you for reading. Thank you for championing these words. Thank you for being my muses. And, in case no one has told you today, I love you. It matters immensely that you're here.

NOTES

Chapter 2: Maps

1. Renée Onque, "Feel Anxious When You Don't Have Your Cell Phone? You May Have 'Nomophobia'—How to Spot the Signs," CNBC, September 2, 2023, https://www.cnbc.com/2023/09/02/nomophobia-the-anxiety-you-feel-without-your-cell-phone-has-a-name.html.
2. David Greenfield, quoted in Catherine Price, "Putting Down Your Phone May Help You Live Longer," *New York Times*, April 24, 2019, https://www.nytimes.com/2019/04/24/well/mind/putting-down-your-phone-may-help-you-live-longer.html.
3. Cleveland Clinic, "Why Multitasking Doesn't Work," *Health Essentials*, March 9, 2021, https://health.clevelandclinic.org/science-clear-multitasking-doesnt-work/.
4. R Blank, "Technology and Information Overload: How Digital Overstimulation from Your Gadgets Harms Your Well-Being," *Healthier Tech Lifestyle*, accessed March 11, 2024, https://www.healthiertech.co/technology-information-overload/.

Chapter 3: Tiny Forks

1. Brianna Wiest, *The Mountain Is You: Transforming Self-Sabotage into Self-Mastery* (Brooklyn, NY: Thought Catalog Books, 2020), 90.

Chapter 4: Checking In

1. Vivek H. Murthy, *Together: The Healing Power of Human Connection in a Sometimes Lonely World* (New York: HarperCollins, 2020), 9.
2. Anne Lamott, *Almost Everything: Notes on Hope* (New York: Riverhead Books, 2018), 67.

Chapter 5: Exactly Where You Need to Be

1. Thomas Merton, *The Sign of Jonas* (New York: Harcourt, 1981), 10.

Chapter 6: A Prayer for Tearing Mushrooms

1. Catherine Doherty, "The Duty of the Moment: Words from Catherine Doherty (1896–1985)," Domestic-Church.com, accessed March 11, 2024, https://domestic-church.com/CONTENT.DCC /19990101/ESSAY/duty_moment.htm.
2. Brother Lawrence, *The Practice of the Presence of God* (New Kensington, PA: Whitaker House, 1982).

Chapter 7: An Ode to Maggie

1. Steve Lohr, "Smartphone Rises Fast from Gadget to Necessity," *New York Times,* June 9, 2009, https://www.nytimes.com/2009 /06/10/technology/10phone.html.
2. Bertha Coombs, "Loneliness Is on the Rise and Younger Workers and Social Media Users Feel It Most, Cigna Survey Finds," CNBC, January 23, 2020, https://www.cnbc.com/2020/01/23/loneliness-is -rising-younger-workers-and-social-media-users-feel-it-most.html.

Chapter 8: I Was Here

1. Kristen Fuller, "How Creating a Sense of Purpose Can Impact Your Mental Health," *Psychology Today,* March 7, 2022, https:// www.psychologytoday.com/us/blog/happiness-is-state-mind/202203 /how-creating-sense-purpose-can-impact-your-mental-health.

Chapter 9: Screen Time

1. Tudor Cibean, "Adults in the US Check Their Phones 352 Times a Day on Average, 4x More Often Than in 2019," Techspot, June

5, 2022, https://www.techspot.com/news/94828-adults-us-check
-their-phones-352-times-day.html.

Chapter 10: Wild and Precious

1. Oliver Burkeman, *Four Thousand Weeks: Time Management for Mortals* (New York: Picador, 2021), 33.
2. Anna Altman, "The Year of Hygge, the Danish Obsession with Getting Cozy," *The New Yorker*, December 18, 2016, https://www.newyorker.com/culture/culture-desk/the-year-of-hygge-the-danish-obsession-with-getting-cozy.
3. Justin Whitmel Earley, *Habits of the Household: Practicing the Story of God in Everyday Family Rhythms* (Grand Rapids, MI: Zondervan, 2021), 4–5.
4. Mary Oliver, "The Summer Day," *New and Selected Poems* (Boston: Beacon, 1992), 94.

Chapter 14: Reframing Productivity

1. Max Fisher, *The Chaos Machine: The Inside Story of How Social Media Rewired Our Minds and Our World* (Boston: Little, Brown and Company, 2022), 30.

Chapter 16: Enjoy the Feast

1. Rebecca Webber, "The Comparison Trap," *Psychology Today*, November 7, 2017, https://www.psychologytoday.com/us/articles/201711/the-comparison-trap.
2. Mattha Busby, "Social Media Copies Gambling Methods 'to Create Psychological Cravings,'" *Guardian*, May 8, 2018, https://www.theguardian.com/technology/2018/may/08/social-media-copies-gambling-methods-to-create-psychological-cravings.

Chapter 18: Status Updates

1. *Merriam-Webster*, s.v. "vulnerable (*adj.*)," accessed November 26, 2023, https://www.merriam-webster.com/dictionary/vulnerable.
2. Vivek H. Murthy, *Together: The Healing Power of Human Connection in a Sometimes Lonely World* (New York: HarperCollins, 2020), 113.

3. Sherry Turkle, "Connected, but Alone?," filmed February 2012 in Long Beach, CA, TED video, 6:54, https://www.ted.com/talks /sherry_turkle_connected_but_alone?language=en.

4. Donald Miller, *Scary Close: Dropping the Act and Finding True Intimacy* (Nashville: Thomas Nelson, 2014), 7.

Chapter 19: Wonder

1. Tonk, comment on "Why Do So Many Adults Lose Their Sense of Wonder?," Reddit, accessed October 16, 2023, https:// www.reddit.com/r/simpleliving/comments/96uxea/why_do_so _many_adults_lose_their_sense_of_wonder/.

2. Summer Allen, "The Science of Awe," Greater Good Science Center at UC Berkeley, September 2018, https://ggsc.berkeley .edu/images/uploads/GGSC-JTF_White_Paper-Awe_FINAL.pdf.

3. Jake Eagle and Michael Amster, *The Power of Awe: Overcome Burnout and Anxiety, Ease Chronic Pain, Find Clarity and Purpose—in Less Than 1 Minute per Day* (New York: Hachette, 2023), 47.

Chapter 21: Rabbi

1. Lauren O'Neill, "Goodbye to the Influencer Decade, and Thanks for Nothing," *Vice*, December 12, 2019, https://www.vice.com/en /article/vb55wa/instagram-influencers-history-2010s.

2. John Mark Comer, "Practicing the Way," November 5, 2021, in *John Mark Comer Teachings*, podcast, https://open.spotify.com /episode/1zHCuL4l1HsxTmoPKxPwk9?si=vxFCw1pXT5mHE2s 9adZt4w&nd=1.

3. Blaise Pascal, quoted in Adam Wernick and Annie Minoff, "A New Study Found People Are Terrible at Sitting Alone with Their Thoughts. How about You?," *The World*, July 19, 2014, https://theworld.org/stories/2014-07-19/new-study-found-people-are -terrible-sitting-alone-their-thoughts-how-about-you.

Chapter 22: Sabbath

1. Abraham Joshua Heschel, *The Sabbath* (New York: Farrar, Straus and Giroux, 1979), 13.

2. Paige Leskin, "Staying Up Late Reading Scary News? There's a Word for That: 'Doomscrolling,'" *Business Insider*, April 19, 2020, https://www.businessinsider.com/doomscrolling-explainer-coronavirus-twitter-scary-news-late-night-reading-2020-4.

Chapter 23: Watch

1. Blue Letter Bible, s.v. *"gregoreo,"* https://www.blueletterbible.org/lexicon/g1127/kjv/tr/0-1/.
2. *Merriam-Webster*, s.v. "vigil (n.)," accessed November 20, 2023, https://www.merriam-webster.com/dictionary/vigil.
3. Heather Hughes, "Keeping Vigil," *Christian Reflection: A Series in Faith and Ethics*, Center for Christian Ethics at Baylor University, 2013, https://ifl.web.baylor.edu/sites/g/files/ecbvkj771/files/2022-12/LentStudyGuide5.pdf.
4. Henri Nouwen, *Following Jesus: Finding Our Way Home in an Age of Anxiety* (New York: Convergent, 2019).

Chapter 25: The Power of Presence

1. Christianity.com editorial staff, "What Does 'Imago Dei' Mean? The Image of God in the Bible," Christianity.com, updated October 21, 2022, https://www.christianity.com/wiki/bible/image-of-god-meaning-imago-dei-in-the-bible.html.

Chapter 26: The Inner Life

1. John Ortberg, *Soul Keeping: Caring for the Most Important Part of You* (Grand Rapids, MI: Zondervan, 2014), 40.
2. Susanne Biro, "Are You Focused on Your Outer Life Or on Your Inner Life?," *Forbes*, April 21, 2021, https://www.forbes.com/sites/forbescoachescouncil/2021/04/12/are-you-focused-on-your-outer-life-or-on-your-inner-life.
3. Gordon MacDonald, *Ordering Your Private World*, rev. ed. (Nashville: W Publishing, 2017), 14.
4. Anne Lamott, *Operating Instructions: A Journal of My Son's First Year* (New York: Anchor Books, 1993), 135.
5. Carol McLeod, *Vibrant: Developing a Deep and Abiding Joy for All Seasons* (New Kensington, PA: Whitaker House, 2020).

Chapter 27: Pace

1. Kosuke Koyama, *Three Mile an Hour God* (London: SCM Press, 2021), 5.

Chapter 28: The Year of the Vegetables

1. John Ortberg, *The Life You've Always Wanted: Spiritual Disciplines for Ordinary People* (Grand Rapids, MI: Zondervan, 2002), 83.

Chapter 29: Before the Noise Got In

1. Manoush Zomorodi, *Bored and Brilliant: How Spacing Out Can Unlock Your Most Productive and Creative Self* (New York: St. Martin's Press, 2017), 5.

Chapter 31: Spacious Places

1. Henri J. M. Nouwen, *Life of the Beloved: Spiritual Living in a Secular World* (New York: Crossroad, 2002).

Chapter 32: Proceed Quietly

1. Sherry Turkle, "Connected, but Alone?," filmed February 2012 in Long Beach, CA, TED video, 13:40, https://www.ted.com/talks /sherry_turkle_connected_but_alone?language=en.

Come Matter Here

Your Invitation to Be Here in a Getting There World

Hannah Brencher

From viral TED Talk speaker and founder of The World Needs More Love Letters, Hannah Brencher's *Come Matter Here* is the book you need to start living like you mean it here and now.

Life is scary. Adulting is hard. When faced with the challenges of building a life of your own, it's all too easy to stake your hope and happiness in "someday." But what if the dotted lines on the map at your feet today mattered just as much as the destination you dream of?

Hannah thought Atlanta was her destination. Yet even after she arrived, she found herself in the same old chase for the next best thing . . . somewhere else. And it left her in a state of anxiety and deep depression.

Our hyperconnected era has led us to believe that life should be a highlight reel—where what matters most is perfect beauty, instant success, and ready applause. Yet as Hannah learned, nothing about faith, relationships, or character is instant. So she took up a new mantra: Be where your feet are. Give yourself a permission slip to stop chasing the next big thing, and come matter here. Engage the process as much as you trust the God who lovingly leads you.

If you're tired of running away from your life or of running ragged toward the next thing you think will make you feel complete, *Come Matter Here* will help you do whatever it takes to show up for the life God has for you. Whether you need to make a brave U-turn, take a bold step forward, or finish the next lap with fresh courage, you will find fuel and inspiration for the journey right here.

Available in stores and online!

Fighting Forward

Your Nitty-Gritty Guide to Beating the Lies That Hold You Back

Hannah Brencher

Find the hope and encouragement you need to overcome anxiety and fear and take the next small step to a better life. Join popular blogger, viral TED Talk speaker, and founder of The World Needs More Love Letters, Hannah Brencher, as she shares personal stories of developing daily rhythms and sustainable faith in a culture of hustle.

At the darkest point of a life-altering depression, Hannah took a silver marker and labeled a composition book with two life-changing words: "Fight Song." In that little notebook, she poured hope-filled truths and affirmations, knowing that one day, she—and you—would need a reminder to stay in the fight. Drawn from those glow-in-the-dark words, *Fighting Forward* is your invitation to show up, claim hope, and take back your life one small win at a time.

With a heap of hope for those who long to move from anxiety and fear into action steps, the power-ballad essays in this book will encourage you to:

- Savor the milestones you've already reached
- Root yourself in the next small step
- Welcome healthy routines into your day
- Apply grace like sunscreen in the process of becoming who you're meant to be

Fighting Forward champions the truth that each song starts with a single note. With trust and a little time, each note and every small step adds up to a victorious anthem of showing up to this life and staying in the fight to become who God made you to be.

Available in stores and online!

From the Publisher

GREAT BOOKS

ARE EVEN BETTER WHEN THEY'RE SHARED!

Help other readers find this one:

- Post a review at your favorite online bookseller

- Post a picture on a social media account and share why you enjoyed it

- Send a note to a friend who would also love it—or better yet, give them a copy

Thanks for reading!